SpringerBriefs on Pioneers in Science and Practice

Texts and Protocols

Volume 3

Series Editor

Hans Günter Brauch

For further volumes:
http://www.springer.com/series/11446
http://www.afes-press-books.de/html/SpringerBriefs_PSP_TP.htm

Rodolfo Stavenhagen

The Emergence of Indigenous Peoples

Springer

EL COLEGIO DE MÉXICO

Rodolfo Stavenhagen
Center for Sociological Studies
El Colegio de Mexico
México, D.F.
Mexico

ISSN 2194-3125 ISSN 2194-3133 (electronic)
ISBN 978-3-642-34143-4 ISBN 978-3-642-34144-1 (eBook)
DOI 10.1007/978-3-642-34144-1
Springer Heidelberg New York Dordrecht London

Library of Congress Control Number: 2012949709

Printed on acid-free paper

Springer is part of Springer Science+Business Media (www.springer.com)

I am pleased to dedicate this book to my numerous indigenous friends in many parts of the world who have taught me the deep sense of their struggles for human rights and justice so long denied

Books by the Same Author Published by Springer

- Rodolfo Stavenhagen: *Pioneer on Indigenous Rights*. Springer Briefs on Pioneers in Science and Practice, vol. 2 – presented by Úrsula Oswald Spring (Heidelberg – Dordrecht – London – New York: Springer-Verlag, 2013).

- Rodolfo Stavenhagen: *Peasants, Culture and Indigenous Peoples: Critical Issues*. Springer Briefs on Pioneers in Science and Practice, vol. 4, Subseries with Texts and Protocols, vol. 1 (Heidelberg – Dordrecht – London – New York: Springer-Verlag, 2013).

The cover photograph of a group of young Masai women in Kenya was taken during a field mission by Rodolfo Stavenhagen as the first United Nations Special Rapporteur on the human rights of indigenous peoples. (From the author's private collection.)

With Mapuche community in Chile at the site of a proposed dam. *Source*: Personal photo collection of the author

Rodolfo Stavenhagen with Sheila Watts Clouthier, Inuit defender of the Arctic Environment, in Nunavut, Canada. *Source*: Personal photo collection of the author

Acknowledgments

The author is exceedingly grateful to Hans Günter Brauch, the most efficient and dynamic editor of these three Springer books, he has prepared on the occasion of my eightieth birthday.

Contents

Chapter 1
Introduction

This is the second of three volumes in this series containing essays written by Rodolfo Stavenhagen over a period of fifty years. In this volume, devoted mainly to the complex topic of the recent emergence of indigenous peoples as new social and political actors on the international scene as well as at the national level in many countries, the author focuses on different perspectives of the social sciences concerning these issues, with Latin America and Mexico as his main geographical interest.

Rodolfo Stavenhagen with President Evo Morales of Bolivia, 2007. *Source* Personal photo collection of the author

R. Stavenhagen, *The Emergence of Indigenous Peoples*, Texts and Protocols 3,
DOI: 10.1007/978-3-642-34144-1_1, © The Author(s) 2013

Rodolfo Stavenhagen at a meeting of the Ayllu Council in Bolivia, 2007. *Source* Personal photo collection of the author

The first chapter, published in Brazil in 1963, critically examines anthropological approaches to the study of Central American indigenous communities that were common at that time, and suggests alternative perspectives. There follows a text on indigenous peoples written many decades later that appeared as an entry in the Oxford Encyclopedia of Human Rights of 2009, and a background paper prepared for the United Nations Human Development Report in 2004. The next three chapters focus on the emerging indigenous movements in Latin America and specifically in Mexico at the turn of the present century.

Chapter 2
Classes, Colonialism and Acculturation (1965)

Essay on a System of Inter-Ethnic Relations in Mesoamerica

Abstract The purpose of this article is to analyze the ethnic relations which characterize the intercultural regions of Altos de Chiapas in Mexico and in Guatemala (This chapter was first published in 1965 as: "Classes, Colonialism and Acculturation", in: *Studies in Comparative International Development*, I,6: 53–77. The permission to reprint this text was granted by the permissions office of Springer in Dordrecht, The Netherlands). It is not my intention to add new data presently unknown to experts in the area. My purpose is both more modest and more ambitious. It is that of reorganizing known data into a scheme of interpretation differing from those which are currently used in anthropology; and which I believe to be more fruitful for the purpose of clarifying some historical and structural problems in the formation of national societies of Mexico and Guatemala (The author expresses his thanks to Guillermo Bonfil, Andrew G. Frank, Carlos Alberto de Medina, and Roberto Cardoso de Oliveira for the comments, criticisms, and suggestions which they have contributed.).

On this particular subject it is my opinion that the conceptual frame of reference of the analysis of social classes is more adequate to the understanding of relationships between economy and society than the frames of reference generally employed by researchers. In the course of the essay I shall use some concepts which are

---

This essay was originally published in Spanish in the journal America Latina, Rio de Janeiro, Centro Latinoamericano de Pesquisas em Ciencias Sociais, 1963, where I was working at the time. It became a chapter of my PhD dissertation at the University of Paris (1965), later published as Social Classes in Agrarian Societies (New York: Doubleday, 1975). In this text, I challenge the then prevailing view, held by many anthropologists, of Indian communities in Mexico and Central America as relatively isolated self-contained ahistorical cultural units outside the national society; a view that supported the assimilationist policies adopted by the state to 'integrate' the 'backward' Indians. Here I develop the concept of 'internal colonialism', as an alternative approach which has been widely used and further developed over the years by social scientists in other countries.

R. Stavenhagen, *The Emergence of Indigenous Peoples*, Texts and Protocols 3,
DOI: 10.1007/978-3-642-34144-1_2, © The Author(s) 2013

sometimes ambiguous. In each case I will try to specify their meaning. But this will not always be possible. In such cases these concepts will have to be understood in their more common sense use. The bibliography cited is merely illustrative. It does not pretend to be exhaustive. Many of the facts analyzed are sufficiently well known so as to require no further documentation. The choice of a region, which includes areas of Mexico and Guatemala, is justified because of cultural and historical similarities of the Indian region on both sides of the border. Political and economic differences between both countries, especially in the course of the last few years, do not seem to have substantially modified the quality of inter-ethnic relationships; particularly is this the case on the analytical level at which this essay is written.

The Maya region of Altos de Chiapas and Guatemala has the peculiarity of each local community constituting a cultural and social unit which is distinguished from other similar communities; and whose limits, furthermore, coincide with those of modern political-administrative units called municipalities or municipal agencies. Thus, the Indian population of every municipality (or municipal agency) can be distinguished from others through their clothing, dialect, membership, and participation in a religious and political structure of their own. This usually involves economic specialization as well; and also a developed feeling of identity with other members of the community, reinforced by a somewhat generalized endogamous system. Aside from being an administrative unit integrated in Mexican and Guatemalan national political structures, the municipality represents in this region the sphere of the Indian population's social unit, which has been called 'tribe' by some ethnologists, and which others have even termed the germ of the 'nation'.[1] This coincidence of modern municipal institutions with traditional Indian structures, resulting from the particular historical evolution of the region, has allowed the survival of the latter within the framework of the modern national state.

2.1 Indians and Ladinos

In the entire region and in almost all of the local communities there co-exist two kinds of populations, two different societies: Indians and Ladinos. The problem of the relationships between these two ethnic groups[2] has been undertaken in

---

[1] Sol Tax, 1937: "The Municipios of the Midwesern Highlands of Guatemala", in; American Anthropologists, 39; Henning Siverts, 1956: "Social and Cultural Changes in a Tzeltal (Mayan) Municipio, Chiapas, Mexico", in: *Proceedings of the 32nd International Congress of Americanists* (Copenhagen).

[2] By *ethnic group* we understand a social group whose members participate in the same culture, who may sometimes be characterized in biological or racial terms, who arse conscious of belonging to such a group and who participate in a system of relations with other similar groups. An *ethnie* may be, depending on circumstances, tribe, race, nationality, minority, caste, cultural component, etc., according to the meaning given to these terms by different authors.

different ways by anthropologists. Only a few of them, nonetheless, have attempted an interpretative analysis within the sphere of the total society.[3] In these pages I intend to offer some elements for such an analysis.

It is a well known fact that biological factors do not account for the differences between the two populations; we are not dealing with two races in the genetic sense of the term. It is true, of course, that in a general way the so-called Indian population answers to biologic traits corresponding to the Amerinds and equally, that the so-called Ladino population shows the biologic traits of the Caucasoids. But even though Ladinos tend to identify with whites, in fact they are generally *mestizo*. It is the social and cultural factors which are taken into account to distinguish one population from the other.

For a long time it was common to draw up a list of identifiable cultural elements in order to distinguish both groups: language, clothing, agricultural technology, food, religious beliefs, etc. The advantages of such a list are that it allows an easy quantification of Indian and Ladino populations, and that census returns which include some of these elements—principally the language—can be profitably used. Thus, using these indices, Whetten was able to speak of the 'indo-colonial' population of Mexico.[4] Confronted with the obvious insufficiency of this procedure in terms of a deeper analysis, it came to be recognized that these cultural elements were integrated within cultural complexes. Alfonso Caso used as his point of departure the fact that Indian populations live in communities which can be easily distinguished from one another, and he thus offered the following definition: "an Indian is he who feels he belongs to an Indian community, and an Indian community is that in which there exists a predominance of non-European somatic elements, where language is preferentially Indian, possessing within its material and spiritual culture a strong proportion of Indian elements and finally, having a social feeling of being an isolated community within surrounding ones, distinguishing it from white and mestizo villages."[5] This definition no longer considers the Indian as an isolated individual, but as a member of a well-defined social group. The author limits the qualification of Indian to a subjective feeling, and introduces racial considerations when distinguishing the Indian community from 'white and mestizo' ones. We do not find in this definition the elements needed for an analysis of the existing relationships between Indians and Ladinos; quite the

---

[3] The global society is the widest operational social unit within which the studied relations take place and which is not a part of the immediate experience of the actors in the social system. It includes the community, the municipality, the region, the ethnic group, etc., and their diverse systems of interrelation. It is sociologically structured. The global society has been ermed a macroscopic group embracing the functional groupings, social classes and conflicting hierarchies. Generally, in this essay, it is identical to the nation (or to the colony), but it sometimes also refers to the wider economic system, in which the nation participates. See Georges Gurvitch, 1950: *La Vocation Actuelle de la Sociologie* (Paris): 301, *passim*.

[4] Nathan Whetten, 1948: *Rural Mexico* (Chicago).

[5] Alfonso Caso, 1948: "Definición del indio y lo indio", in: *América Indigena*, 8,5.

opposite, Caso's definition stresses the idea that we are dealing with two autonomous cultural worlds whose co-existence is almost a matter of chance.

The importance attributed by ethnologists to cultural elements of Indian populations has long concealed the nature of socio-economic structures into which these populations are integrated. Sol Tax, for instance, while studying Indian economy in Guatemala, chooses a community in which one-third of the population is Ladino. Yet Tax describes only the Indian aspect and leaves aside the *mestizo* population as though the community's economy was not a complex and integrated whole. When he is forced to describe the inevitable interaction taking place between Indians and Ladinos, he does so as though he were dealing with external relations of Indian society.[6] Siverts, when speaking about monetary exchanges between Indians and Ladinos, even uses the term 'external commerce.'[7]

Certain recent ethnographic studies, and primarily the needs of *indianist* activity in Mexico, have shown the weaknesses of an approach based exclusively upon analysis of cultural factors, not taking into account historical evolution. Eric Wolf has recently declared that "the condition of the Indian does not consist in a discreet list of social traits; it lies in the quality of social relationships found among communities of a certain kind and in the self-image of the individuals who identify with those communities. The Indian condition is also a distinctive historical process, since these communities originate at a given moment, grow stronger, decline again, and maintain or lose stability in the face of attacks or pressures coming from the larger society."[8] Thus, it is no longer the cultural patterns but the community structure, the relationships between its different parts, which are significant. The Indian condition is to be found in those closed 'corporate' communities, whose members are bound by certain rights and duties, having their own forms of social control, particular political and religious hierarchies, etc. According to Wolf, these corporate units are the result of Spanish colonial policy, having suffered successive transformations under the impact of external influences. Wolf admits that these units, which are neither totally isolated nor completely self-sufficient, take part in wider economic and political power structures. The Indian communities are related to national institutions and include groups oriented toward both the community and the nation. These groups perform roles as political 'power brokers' between traditional and national structures.[9]

Wolf's analysis of the Indian supplies historical depth and structural orientation which are not found among specialists in cultural anthropology. However, while he clearly recognizes the existence of the corporate community's external

---

[6] Sol Tax, 1953: *Penny Capitalism, A Guatemala Indian Economy* (Washington).

[7] *Loc. cit.*, p. 183.

[8] Eric Wolf, 1960: "The Indian in Mexican Society", in: *Alpha Kappa Delta*, 30,1.

[9] Eric Wolf, 1956: "Aspects of Group Relations in a complex Society: Mexico", in: *American-Anthropologist*, 58.

relations, it seems to respond mechanically to impulses originated in national and regional sources of power. Wolf does not speak about the relationships between Indians and Ladinos. Tax and Redfield also admit the existence of external relations, with the difference that for them, the controls imposed upon the population from outside the local community "have their origin in natural law"![10]

Indianist action in Mexico has forced ethnologists to restate the problem in different terms. There has been a shift from the sphere of the Indian community to that of the intercultural region where Indians and mestizos co-exist. This region possesses the characteristic of having an urban complex mainly inhabited by a Ladino population and surrounded by Indian communities which are its economic and political satellites.[11] This new focus allows a better analysis of socio-economic structures and of relationships between human groups. We no longer speak of acculturation alone, but of the Indian's integration to the nation, which is precisely the stated purpose of Indianist policy. The ecological relationships between the metropolis and its satellites are only a part of the complex system of social relationships characteristic of this region. The theoretical framework used until now in the study of these relationships has proved insufficient for their full interpretation.

2.2 The Land and Social Relations

Class relationships in any society appear only through the analysis of the whole socio-economic structure. In the Indian region of Chiapas and Guatemala these relationships are not visible through the study of cultural differences between the two ethnic groups, nor do they show in all of the social situations in which there are inter-group relations. Class relationships emerge clearly through distribution of land as a means of the labor, trade and property relations which link one part of the population to another.[12]

---

[10] Robert Redfield and Sol Tax, 1952: "General Characteristdics of Present Day MesoamericanIndian Society", in: *Heritage of Conquest* (Glencoe:the Free Press).

[11] Alfonso Caso, 1957: "Los fines de la acción indigenistaen México", in: *Revista Internacional del Trabajo*, December, and G. Aguirre Beltrán, 1957: *El proceso de aculturación* (Mexico: UNAM), which stillconstitutes the most complete theoretical exposition on Mexican nativism.

[12] I use here the terms 'class', 'class relations', and 'class situation' as analytical concepts and I completely distinguish them, as shall be seen later, from the concept of social stratification generally associated with them. For theoretical justification of this methodological procedure see my article on "Estratificación y Estructura de Clases", in: *Ciencias Politicas y Sociales* (Mexico), No. 27 (1962), and my paper on "as relaciones entre la estratificación y la dinámica de clases", presented at the Seminario sobre Estrateficación y Movilidad Social, Rio de Janeiro, 1962 (which shall be published by the Pan American Union.

2.2.1 Production Relations

2.2.1.1 Subsistence Agriculture

The basis of regional production is agriculture, and the basis of agriculture is maize, principally for domestic consumption. Even when other crops are cultivated, maize is the primary agricultural product without which the rural family, the productive unit, would not survive. The soil is poor, agricultural techniques are primitive, and yields are therefore small. Rainfall allows two harvests a year in some regions. The farmer devotes a great part of his time to subsistence farming with participation of family labor. Produce is consumed by the family. Sometimes, when the farmer needs money, he sells part of the harvest, but later, when his reserves are exhausted, he must buy his corn back again. In his position as a maize producer, the farmer remains isolated and does not enter into relations with other sectors of society.

There are exceptions to this situation. Some communities in the area have become specialized in maize production to the exclusion of any other important agricultural activity. Santiago Chimaltenango, in Guatemala, regularly produces a surplus of maize which is sold at the local markets.[13] In this case, the subsistence farmer becomes, in part, a peasant producing for the market. I say in part because due to the fact that the bulk of his production is consumed at home, he remains within a subsistence economy. It is important to stress the fact that maize is grown almost exclusively by the Indians. Even though the majority of the communities have also a Ladino population, these rarely grow maize. When they devote themselves to agriculture, it is usually to produce cash crops.

We find a primary element for differentiation of the population into social classes: one part of the population predominantly devotes itself to subsistence maize farming—even while it sells some surplus—and another sector does not participate in subsistence agriculture.

2.2.1.2 Commercial Agriculture

Almost all of the rural communities also participate in agricultural activities whose purpose is not domestic use but commerce. The subsistence farmer is also a producer for the market. Even while he may not devote the greater part of his time to this activity, it allows him to obtain the money he needs. At altitudes lower than 5,000 feet, maize economy is complemented with that of coffee, a cash crop par excellence. There is also cacao, onion, and vegetables of all kinds. At higher altitudes there are fruits. All of these food products are destined for sale, and the different communities specialize in production of one or the other. Maize and coffee (within their geographic limits) are found everywhere. Coffee is destined to

[13] Charles Wagley, 1957: *Santiago Chimaltenango* (Guatemala).

national and international markets, while the majority of the other products appear only in local markets. The coffee-growing communities are usually richer than those which, located on higher and poorer lands, do not grow it. The subsistence farmer who grows coffee and other products for the market does not neglect growing his maize. Every community, in fact, possesses lands which are used only for maize, and other, usually better, lands which are used for cash crops. The subsistence farmer secures his maize crop first; only if he has time and additional land at his disposal does he devote himself to commercial farming, even if the latter be more productive than the former. In Panajachel, Guatemala, for instance, growing coffee and onions pays better than growing maize. Yet the Indians do not devote themselves to these activities until they have prepared their maize plots.[14] It is obvious that agricultural factors are involved in this situation; in the poorer and more inaccessible soils only maize can be grown, while the flat and fertile soils, which are nearer to the village, are taken up by commercial agriculture. But there are also reasons of an economic variety: the subsistence farmer has to secure his maize first, because he cannot buy it elsewhere. Panajachel produces only little economic surplus, and should the farmer devote himself exclusively to cash crops, without having the possibility of importing maize from outside, the basis of his economy would crumble. We are therefore not dealing with the individual producer's choice alone, but with a problem of economic development.

Besides corn, in this community Indians are able to grow vegetables and coffee. Yet they grow especially the vegetables, notwithstanding the fact that these pay less than coffee. Coffee is a perennial plant, and the establishment of plantations requires time and capital. Since the Indians lack the means, they prefer to grow vegetables, with which they are able to obtain quicker, if smaller, benefits. Sol Tax describes the Panajachel Indians' economy as being a 'penny capitalism', because they produce commercial agriculture for the market, because they are oriented toward a profit economy, and because they like to make 'a good deal'. Nonetheless, Tax himself shows that their economy is dominated in the first place by the needs of maize farming, and that they prefer to grow vegetables rather than coffee, although coffee pays more. The reason for this apparent contradiction lies in the fact that the Indians lack capital and credit institutions. As Wolf has pointed out,[15] it is precisely these two factors—non-existent in Panajachel—which define a capitalist system. The Panajachel Indian is integrated to the capitalist system, through the sale of his coffee and acquisition of industrial products. But the subsistence farmer, the Indian, is not the 'capitalist' in this case. On the contrary, he is placed at the opposite pole. His agricultural labor is not essentially a commodity, and the money he earns through the sale of his vegetables is not reinvested but spent in current consumption. There is no accumulation of capital.

In differing from the Indians, Ladinos do not grow maize but only cash crops. They settled in the region in the course of the past century, with the expansion of

[14] Sol Tax: *Penny Capitalism, op. cit.*

[15] Eric Wolf, "The Indian in Mexican Society", *loc. cit.*

coffee. In the rural communities the Ladino farmers are few in number, and farming is never their only occupation. In Panajachel, they grow the greater part of the coffee, and their farming is exclusively commercial. The coffee producer always employs salaried labor; he therefore has the necessary capital available. He is, in fact, a capitalist farmer, and he is able to afford it because, differing from the Indian, he does not devote his time to subsistence farming. The growing of coffee, as well as those who grow it, were introduced from outside. The Indians have accepted this new kind of farming only as a complementary economic activity.

Here we have a second element for the differentiation of social classes. We distinguish on the one hand, the farmer devoted to commercial agriculture as a complementary activity, and who obtains from it only minimal profits which are wholly destined for consumption; and on the other, the farmer (especially the coffee-grower) who accumulates capital, employs labor, and who usually also performs other non-agricultural activities. Again, the former are Indians and the latter Ladinos.

2.2.1.3 The Agricultural Workers

Until now we have spoken only about independent farmers, but a large part of the farming population is composed of laborers. In Jilotepeque (Guatemala), laborers constitute 90 % of the active population, of which only 9 % are Ladinos. All of the laborers work for Ladinos; there is not one Indian in this community who employs labor.[16] In the highlands of Chiapas, the peasants regularly work as laborers in the big coffee plantations, where they spend many months a year. Till only recently, this was forced or semiforced labor, and the contract and employment conditions were notoriously bad. At present there exist labor unions of Indian workers, and the Mexican government has taken measures for the protection of migrant workers. Nonetheless, recruitment of laborers is still done by pressures and coercion which sometimes exceeds the legal limits of what is called a free contract. From an Indian population totaling 125,000 persons in this area of Chiapas, 15,000 laborers are employed on a seasonal basis.[17] In Guatemala's coffee plantations compulsory labor for Indians existed until recently; up to a maximum of 150 days per year, depending upon the amount of land which they possessed. The pretext for this recruitment was the fight against idleness; yet no Ladino, even those possessing no lands, was forced to perform this kind of work.

It is obvious that the laborer is placed in a class situation. This is perhaps more so for those who emigrate temporarily from their communities in order to work in the plantations than it is for those who remain at home and work as laborers in plantations closer to their communities. These laborers are not separated from the

[16] Melvin Tumin, 1952: *Caste in a Pesant Society* (Princeton).

[17] A.D. Marroquín, 1956: "Consideraciones sobre el problema de la region tzeltal-tztzil", in: *América Indígena*, 16,3.

social structure to which they belong; they remain subsistence farmers. They go in search of salaried work only when their corn field is secure. Writing about the chamulas, Pozas says that they do not want to work in coffee plantations, and that they do so only when compelled by economic needs.[18] In Guatemala, temporary migrations in search of work annually affect 200,000 Indians,[19] and more than one half of the big plantations' laborers are migratory. "This recruitment", one author says, "has been the means by which the plantations have extended their influence over almost all Indian communities in Guatemala".[20]

Insofar as the monetary needs of these rural communities are concerned, salaried labor has in some of them the same economic function as commercial agriculture has in others. From the point of view of the global economic structure, the self-subsisting community functions as a labor force reserve.[21] The degree of economic exploitation inflicted upon this labor force is shown by the following datum: in Jilotepeque, a Ladino laborer earns 50 % more than an Indian laborer, yet the cost of supporting a mule is even higher than a Ladino's salary![22]

It can thus be seen that salaried work and commerce notwithstanding, the structure of self-subsisting communities has not been wholly broken down. In Cantel, a Guatemalan community, only when the farmer does not possess enough land to feed his family does he seek work in a textile factory located there. The industrial worker remains integrated in the structure and values of his community. The new class relationships produced by local industrialization have only partially modified traditional structure. Here industrial work has the same function as migratory work and commercial agriculture in other communities.[23]

Salaried work represents a third element in terms of class differentiation in the area. The monetary income obtained by farmers in the manner described above represents the complement to a subsistence economy. We find here new production relations, in which the Indian is always the employee and the Ladino the employer. When there are Ladinos employed by other Ladinos, they occupy higher positions and receive higher salaries than the Indians.

We are now ready to attempt a first generalization. At the level of agricultural production, the relationships between Ladinos and Indians are class relationships.

---

[18] R. Pozas, 1959: *Chamula, un pueblo de los Altos Chiapas* (Mexico).

[19] M. Monteforte Toledo, 1959: *Guatemala, monografía sociológica* (Mexico).

[20] A. Y. Dessaint, 1962: "Effects of the Hacienda and Plantation Systems on Guatemala's Indians", in: *América Indígena*, 22,4.

[21] Dessaint, *loc. cit., writes*: to "obtain adequate supply of labor has always been of basic importance ever since the Spanish Conquest" (p. 326). And Oliver La Farge has said: "Two methods have been used to tap the great source of labor of the highlands: violence and the destruction of the economic bases which allowed the Indians to refuse voluntary work in the lowlands." ("Etnología maya: secuencia de culturas", in: *Cultura Indígena de Guatemala* (Guatemala, 1959).

[22] Melvin Tumin, *op. cit.*

[23] *Cf.* Manning Nash, 1958: *Machine Age Maya: the Industrialization of a Guatemalan Community* (Glencoe).

The former produce exclusively for the market, while the latter produce primarily for their own consumption; Ladinos accumulate capital, Indians sell their farming products only in order to buy goods for consumption; Ladinos are employers and Indians are laborers. These relationships shall be seen with greater clarity when we consider land tenure.

2.2.2 Land Tenure

2.2.2.1 Communal Property

The system of land ownership in colonial times worked against Indian lands. Through grants and patronage, Indian communities were deprived of their lands. The tutelary legislation of Indians, which protected communal property, was difficult to apply in practice. During the national period the collective lands survived only in the more isolated regions of New Spain, such as the one we are now discussing. The liberal reforms of the past century were equally directed against communal property. Part of the population, nonetheless, still possesses communal lands to the present day. There are various forms of collective land tenure, and their legal aspects are not always clear. Sometimes these are lands which in effect belong to a community, in accordance with a land title of the Colonial period, having been revalidated once in a while by some later national government. Another variant is that in which the deed of land ownership is held by an elder of the community who in fact is no more than a trustee. There is no precise data on the subject, yet it seems that traditional kinds of communal lands are not very common in the area. A survey carried out in 80 different villages of West Guatemala showed the existence of communal lands in only one community.[24] In Mexico, the agrarian reform has modified the nature of collective lands in a great number of communities.

The still existing collective property is generally composed of poor soils, hardly useful for farming, and of minimal productive and commercial value. These lands are generally used for pasture, for gathering wood and wild fruit. All members of the community have a right to use these lands. Sometimes communal lands are also used to grow corn. In communities where this is done, the extent of communal lands is never sufficient to satisfy all of the farmers' needs. Thus, it can only absorb a part of the farming labor. Only very rarely are communal lands used for commercial farming purposes, and when such is the case, the monetary economy exerts a pressure upon maintenance of collective property. Tax cites the case of some fruit trees planted on communal lands of a Guatemalan village, which are the

---

[24] *Cf.* Goubaud's remarks in the discussion of the report by Sol Tax: "Economy and Technology", in: S. Tax (ed.): *Heritage of Conquest*, op. cit., p. 74.

object of commercial transactions even while the land is still indivisible.[25] In a Chiapas community, the Indians collectively bought an estate which has now been integrated to the communal possessions of the lineage[26]; but usually communal lands are very ancient.

A community still possessing communal lands is also a traditional community, relatively well integrated from a social point of view and more or less homogeneous from an ethnic point of view. If land cannot be sold, it is unlikely that Ladinos will be allowed to use it. It is also a poor community, with an economy for subsistence, since fertile soils and the possibilities of commercial agriculture attract the Ladinos and tend to transform collective property into private property. In other words, traditional collective lands are infrequent and do not perform an important role in the economy and social organization of Indian communities of this region.

2.2.2.2 The Ejido (Public Land)

Agrarian reform in Mexico reached the Indian region of Chiapas during the regime of President Cardenas. In some communities traditional collective lands were transformed into *ejidos*; in others, some of the latifundia were expropriated in behalf of the peasants. In general, the distribution of ejidos respected ethnic differences, so that each ejido includes in effect members of an homogeneous and socially integrated ethnic group, which accentuates its character of being communal property. The proportion of ejido lands varies with respect to total property in the different municipalities. In ten municipalities, in which the density of Indian population is very high, ejido property is distributed in the following way: in three municipalities it embraces almost 100 % of the total number of properties. Here we are obviously dealing with traditional communal lands which have resisted the process of disorganization characteristic of other communities, and which are now protected by agrarian legislation through legally sanctioned land tenure. In two municipalities ejido property represents more than 65 %; in yet two others, more than 35 %; and in the remaining three, less than 25 %. Thus in the region there is no general tendency with respect to the proportion of ejidal lands.[27]

In Guatemala the existence of ejido lands may be considered as a tenacious defense of traditional Indian collectivities against the economic system represented by private property and by the Ladino group. In Mexico, on the contrary, the ejido is the result of an active struggle for the land by the Indians against the great latifundists. This struggle, which has had its history of violence, is already an old one, and was recently stimulated by the national movements of agrarian reform. Here, as in other parts of Indian America, the agrarian struggle has often taken the

---

[25] *Penny Capiralism, op. cit.*

[26] Calixta Guiteras Holmes, 1961: *Perils of the Soul* (Glencoe).

[27] Ricardo Pozas, 1959: *Chamula, un pueblo indio de los Altos de Chiapas* (Mexico).

shape of an inter-ethnic conflict. Yet at the same time it is an agent of acculturation, despite the apparently contradicting fact that its manifest objective is the reconstitution of the traditional Indian communities' territorial base.

Despite the fact of being collective property, ejidal lands are tilled individually, or rather by the family group. In Chamula, where all of the land is ejidal, the families control their plots as though they were private property, yet without being able to subdivide them. These plots can be inherited by sons and daughters alike, and this has produced a progressive atomization of family 'property', the result of which has been the emigration of a large number of Chamulas in search of lands in the neighboring municipalities. In other communities, the farmer is entitled to the use of ejido lands only as long as he regularly works them. This condition is characteristic of traditional communal organization and follows the Mexican national agrarian reform legislation.

2.2.2.3 Private Ownership of Land

This is the more usual form of land tenure. It was introduced by the Spaniards and spread gready after the nineteenth century's liberal reforms. Under the new liberal legislation Indian communities were forced to transform their communal lands into individual property, which contributed to the fact that many communities completely lost their lands.

Private property of land means that this has an economic value and that it has been transformed into a commodity. It also means the emergence of inequalities among men, according to the extension of the lands they possess, and new social relationships, the basis of which is private property of land: sharecropping, tenant farming, wage labor, sale, mortgage, etc. In Panajachel—writes Tax—the land is fully integrated in the commercial cycles which characterize 'penny capitalism'. But the process is not yet finished. Tax admits that in this community the lands are not considered as an investment (that is, as capital) but only as consumption goods. In Chamula, as we have seen, the land is collectively owned (ejido), yet the concept of private property (even without its juridical manifestations) is developing. The land can be inherited and divided, but not sold. It does not produce rent, but it can be mortgaged under certain special conditions.

In the Indian area, the private property of land has stimulated Ladino penetration. First attracted by the new coffee crop, during the past century, they later took to other kinds of commercial agriculture. Freeing the land in fact accelerated the expansion of the national commercial-capitalist system. In Jilotepeque, Eastern Guatemala, the Indians have progressively lost their lands to such a degree that now only 5 % of the Indians possess enough land to satisfy their needs, while 95 % of them must rent theirs from the Ladinos. 70 % of the land belongs to the Ladinos, who represent only 30 % of the population; and this land is primarily tilled by sharecropper or salaried Indians. The Ladinos possess, as an average, 57.3 acres of land, and the Indians 13.2 acres. The results of a survey showed that among the Indians 16 % of them were landowners, while among the Ladinos,

55 % were.[28] In Panajachel, West Guatemala, the Ladinos represent one-third of the population, but they possess 80 % of the lands. The average Ladino possesses more than eight times more land than the average Indian. Besides, the Ladino often possesses lands in other municipalities.[29] How did it come about that the Ladinos have been able to take possession of such a large amount of land? Charles Wagley tells us: "The inevitable result of the series of laws extolling private property in compliance with modern conceptions was that many Indians who were unable to seize the meaning of the new private documents failed to register their lands, and these were often sold to the big plantations as non-validated lands."[30] Pozas quotes the case of a Governor of the State of Chiapas who, as a result of the Reform laws 'denounced' the existence of communal land in an Indian municipality, and thus obtained legal tide to it. In many instances the Indians' property titles soon passed into the hands of the latifundists, and even when no legal changes in land tenure occurred, the Indians were progressively dispossessed of their land. The lack of land forced the Indians into becoming peons on the big plantations. Many independent farmers were thus depressed to the condition of semi-serfs; and others were recruited for temporary forced labor.[31] This situation was consolidated at the end of the nineteenth century with the political victory of the conservative forces in Mexico and Guatemala.

These examples show that the private property of land benefits the Ladinos and harms the Indians. The process of appropriation of the land by the Ladino element is a unilateral one; it does not work in the opposite direction.[32] In Mexico, nonetheless, it has been possible to check it somewhat, due to the agrarian reform and the ejido system.

Ultimately there exists a great difference between the Ladinos and the Indians in terms of land property, particularly with respect to their use of it and the feelings and attitudes assumed with respect to it. The Indian is a man who is integrated in his traditional community, which is bound to the land. The Indian tills the soil; culturally and psychologically he ceases to be an Indian when he becomes separated from the land. The tilling of the soil is intimately related to the group's social organization (lineage or tribe), and to religious organization and belief. The Indian needs the land because without it he loses his social and ethnic identity. It does not matter whether this land is communal, ejido, or private. In any case, it will be

---

28 Melvin Tumin, 1958: *Caste in a Peasant Society, op. cit.*; John Gillin, 1958: *San Luis Jilotepeque* (Guatemala).

29 Sol Tax: *Penny Capitalism, op. cit.*

30 Charles Wagley: *Santiago Chimaltenango, op. cit.*, p. 67.

31 *Cf.* Calixta Guiteras Holmes, 1961: *Perils of the Soul* (Glencoe), who writes: "In the course of the years more than half the land of the Pedrano Indinas were bought by rich and influential foreigners. … The man who bought the land acquired the right to exploit its occupants" (p. 14). "In 1910 the Indians had not only lost their ownlands but had also become peons" (p. 16).

32 One rare exception to this historical trend in the Guatemalan village of Chitatul, quoted by Richard Adams in his: *Encuesta sobre la cultura de los Ladinos en Guatemala* (Guatemala: EMEP, 1956).

property and not merchandise. It is a means of production, but it is not capital. It is a source of income, but not of rent. Traditionally, the land is not an exchange value for the Indian. The soil must be tilled, and only by doing so does the Indian come to realize himself (even when it be on someone else's property, as day laborer, sharecropper or tenant). The tilling of the land is primarily performed by the family, yet should the need arise a few day laborers may be temporarily employed to help in the farming tasks. The Indians do not like to sell their lands, particularly to the Ladinos; yet throughout the years they have done so when they had to. On the other hand, when land is scarce, as in Chamula, those who are the most dynamic or the most needy go in search of land in other places; either to buy it or to work on communal lands of other municipalities. But they do not break their social ties with their group of origin.

The private property of land is only one aspect of the deep transformations which have affected the Indian communities since the nineteenth century, and which have accelerated during the last decades. Pozas points to the growing contradiction in Chamula between the new principle of private property and the traditional principle of communal and clan equality.[33] This contradiction is not equally profound in other municipalities. In Panajachel, on the contrary, the land is subject to an active commerce among the Indians. Yet he who sells his land loses prestige, while he who buys it, increases it.[34] Also in Chimaltenango, it is disapproved of when the Indians sell their lands, and yet "the lands change ownership with a certain frequency" and there are some Indians who have rather large properties.[35]

From the above we can see that among the Indians private ownership of land is still in a period of transition. For the majority of the Indians, who participate in a communal subsistence economy, land as a means of production has not yet acquired the characteristics which it has in a more highly developed economy. The land is still too much linked to the Indian's socio-religious and family complexes to have become a commodity, an object of a distinctly commercial value which it has become among the Ladinos. Finally, as a juridical instrument, the private ownership of this Indian land has not only failed to provide the Indians with the equality and the security which it was meant to provide, according to the liberal ideology, but quite to the contrary, it has exposed the (relative) independence of these populations to the acquisitive spirit of those representing the new economic structure, the Ladinos.

For Ladinos, the private ownership of land has a different meaning than it has for the Indians. It is associated with commercial farming (especially coffee), with a monetary economy, with wage labor, including a type of servitude of the Indians and, finally, with prestige and personal power. For the Ladinos land is a commercial value, independent of the group's social organization. The Ladinos'

---

[33] *Op. cit.*, p.63.

[34] Sol Tax: *Penny Capitalism, op.cit.*

[35] Charles Wagley, *op. cit,* p. 73, *passim.*

primary goal is to accumulate land and to exploit it through the use of wage labor. The Ladino still has, in part, the aspirations of a feudal lord (the New World variant), but there are very few of them who achieve the privileged position of a big landowner, a position reserved to the descendants of the original owners during the Colonial and post-Colonial period. The Ladino is contemptuous of manual labor; his property serves the purpose of obtaining an income which allows him to devote himself to commerce and politics. Ladinos have not yet acquired a capitalist spirit in the Weberian sense of the word. The development of a regional economy compels him to be an entrepreneur. We have already seen that the majority of the lands belonging at present to the Ladinos were obtained by them at the time of the coffee boom, during the past century. Ladinos use their accumulation of lands to obtain and control cheap labor. The *Instituto National Indigenista* in Mexico has declared that: "In *Altos de Chiapas* diverse *tzeltal* and *tzotzil* communities have seen their lands invaded by neighboring ranchers. Since it is an over-populated region, the land has gradually been impoverished by the long-standing of cultivation, as well as by backward agricultural practices which erode it, and by over-pasturing. With the occupation of their best lands, the Indians find themselves driven each season to the coffee plantations of Soconusco, or working on the margins of the Grijalva under the sharecropping system, subject to the cruel conditions imposed by the owner."[36] Pozas describes the case of a coffee plantation owner who bought a property in an Indian municipality, and who allowed the Indians to grow their corn there under the condition that they would regularly work on his coffee plantation which lay in another region.

This brief analysis has shown that the private ownership of the land has different economic and social functions among the Indians and the Ladinos. It is a social institution linked to the capitalist development of the region. But it primarily benefits the Ladino group, and it is used by them as an instrument of exploitation of the Indians. The private ownership of land, introduced by the liberal regimes who, ironically, wanted the greatest good for the greatest number, has only served to dispossess the Indians of their lands, thus forcing them to go in search of wage work. The private ownership of land thus constitutes one more element for the differentiation of the social classes of the region.

There are also important differences inside the owners' group, of course, but we do not have the data which would enable us to study them in relation to ethnic differences. The Ladino owners generally possess more lands than the Indian owners. Yet in each of these ethnic groups the extension of properties varies a great deal. Minifundists are many in number, and latifundia, though small in number, concentrate the greatest part of private lands. The great latifundists are always Ladinos, of course, and the Indians concentrate at the base of the pyramid. But there are also Ladinos who own only very small parcels of land, while, on the other hand, there are Indians who possess, as in Chimaltenango, 50 times more

---

[36] "La situación agraria de las comunidades indígenas", in: *Acción Indigenista*, No. 105, March 1962.

land than others. The greater part of Indian owners do not possess enough land in order to meet their basic needs, and there are those who sell their minute properties and become day laborers in order to earn a little more.[37]

2.2.3 Commercial Relationships

The Indian economic world is by no means closed. Indian communities are only isolated in appearance. They participate in regional systems and the national economy. Markets and commercial relationships represent the primary link between the Indian community and the Ladino world, between subsistence economy and national economy. It is true that the major part of the Indians' agricultural produce is consumed by them. It is also true that the income generated by the Indians only represents a minimal part of the GNP (even in Guatemala, where the Indian population represents more than one half of the total). The importance of these relationships does not lie in the amount of commercialized products or in the value of the products being bought; it lies in the quality of commercial relationships. These are relationships which have transformed the Indians into a 'minority'[38] and which have placed them in the condition of dependence in which they now find themselves.

Markets and commerce in the region have their background in the pre-Hispanic and colonial period. Their importance in some places is such that Redfield even speaks of a 'primitive merchant society'.[39] Tax calls the system 'capitalist' because it rests on a "monetary economy organized around single households which are units of production and consumption, with a strongly developed market which tends to be perfectly competitive".[40] Such does not seem to be the case in

[37] When general considerations are made on the Mayan case in Chiapas and Guatemala, certain local aspects and particular situations of great interest are necessarily neglected, the inclusion of which would perhaps modify the general scheme. It is a risk of which the author is wholly conscious, yet which he had to assume, considering the limits imposed by an article. Such is the case, for instance, of the Agrarian Reform in Guatemala, initiated with the revolution of 1944, but checked and diverted by the governments subsequent to the 1954 counter-revolution. Thus, the redistribution of the lands, the law of compulsory renting and the constitution of rural workers'labor unions during the decade of 1944-54 surely affected, in diverse ways, the class relations here analyzed. Yet as the processes are no longer in force, I have chosen to ignore them, at the risk of neglecting some facts which might be important to this analysis.

[38] In the sense given to this sociological term by Charles Wagley and Marvin Harris in their *Minorities in the New World* (New York, 1958).

[39] Robert Redfield, 1939: "Primitive Merchants of Guatemala", in: *The Quarterly Journal of Inter-American Relations*, 1,4.

[40] Sol Tax, *Penny Capitalism, op. cit.*, p. 13.

other areas of the region, where the Indian market shows strongly marked monopolistic elements.[41]

Indian markets and the 'constellation of regional markets' have been described in many contexts (especially in Mexico). Thus, it should be unnecessary to offer a detailed analysis of their structure. The role of the Ladino city as a metropolis or urban complex of an intercultural region, and its position of economic, political, social, and religious dominance with respect to satellite Indian communities is very well known. Between the city and the communities there develops a network of close and complex commercial relationships. In the city there is a weekly market of regional importance, and regular and permanent commerce in the stores and in the daily market. At the weekly market place there is an influx of thousands of regional Indians who go to the market to sell their handicraft and farm products, and to buy industrial and handicraft goods at the commercial establishments of the city. Some Indians are full-time traders who participate in the cycle of regional markets; Redfield has called them 'primitive merchants'.[42] But the majority of Indian producers carry their products to the market themselves, usually accompanied by their families. Commerce at the regional urban complex is so organized that the Indian always leaves behind his small monetary income. He sells cheaply and must buy dearly. The Ladino trader perceives a double benefit, through buying the Indian's products and selling him the articles which the Indian family needs not only to satisfy its daily wants, but also those which are related to political and religious life.

Despite Tax's findings in Panajachel, there seems to be a general tendency towards a monopsonic structure in the Indian markets, in which the Indian producer–seller is in no way able to influence the price level. Trading of food products (the basis of Indian production) is controlled by a few Ladino monopolists from the city. As Marroquin has pointed out, the well known bargaining of Indian markets is an instrument used by Ladinos in order to depress price levels of Indian products. In San Cristobal de las Casas, for instance, the same effect is achieved through the performance of the atajadoras, the Ladino women who place themselves at the city's entrance on market days and almost violently force the submissive, incoming Indians to sell them their wares at prices that they impose and which are lower than those which prevail at the market. These varied forms of exploitation which victimize the Indian trader, both as seller and buyer, are due to economic and political dominance of the urban Ladinos. This power is reinforced by their cultural superiority as expressed by their knowledge of price—building mechanisms, of the laws of the country; above all, of the Spanish language, which, being unknown to the Indians, represents one more factor of inferiority and social oppression. It is obvious that under these conditions the Indian has no access to national legal institutions which protect his individual rights.

---

[41] A. Marroquín, 1957: "Introducción al Mercado indígena mexicano", in: *Ciencias Políticas y Sociales*, 8.

[42] Robert Redfield, *op. cit.*

Not only in the city but also in the 'satellite communities' is commerce usually in Ladino hands. The latter are also moneylenders, which is an important function in societies where there is no accumulation of capital and where political and religious life demands considerable expenses. In order to pay their debts, Indians often mortgage their harvest (but seldom their property) and go to work on the coffee plantations.

Among the different kinds of relationships which take place between Indians and Ladinos, commercial relationships are the most important. The Indian participates in these relationships as producer and consumer; the Ladino is always the trader, the middleman, the creditor. The majority of the Indians enter into economic and social relationships with Ladinos at the level of commercial activity, and not at the level of wage labor. It is precisely the commercial relationships which link the Indian world to the socio-economic region in which it is integrated, and to national society as well as to the world economy.

Often commercial relationships go together with social relationships of another kind. Pozas writes that these are sometimes familial. He says that "interdependence between Indian and Ladino individuals and families constitutes the real basis of relationships between the Ladino urban complex and the Indian rural villages".[43] These relationships between families can take the form of compadrazgo (Godfather complex). Although at first sight compadrazgo may appear to be an institution in which Indians and Ladinos face each other on a level of equality, in fact it contributes to accentuate the Indians' condition of inferiority and dependence. *Compadrazgo* is one among many institutions in a complex system which keeps the Indian subordinated to the Ladino in all aspects of social and economic life.

The conjunction of ill these commercial relationships allows us to carry our analysis further. It is obvious that Indian communities are not economically closed. On the contrary, they are linked to regional structures by means of which they participate in the national and world economy. They are the weakest link of a national economy. On the other hand, these commercial relationships are only a part of the Indian community's economic system. It is precisely this one aspect of all the economic activities of Indian communities which places them in a specific and special situation with respect to the Ladino population: a class situation. Commercial relationships between Indians and Ladinos are not relations between equals. The Indian, as a small producer, small seller, small buyer, and finally as a small consumer, can influence neither prices nor market tendencies. The Ladino, on the contrary, holds a privileged situation in the region. The Ladinos, small in number are for the greatest part traders and middlemen. The city, populated by Ladinos, is monopolistic. Regional production is concentrated in it. There finished goods are distributed. True, these activities are a function of regional cities throughout the world. But here the economic inequalities between the city and the community are accentuated by the low level of agricultural production, the high

[43] Ricardo Pozas, *Chamula, op. cit.*, p. 111.

cost of goods brought from other regions, and by all the other means of political, religious, and social power which the city exerts over the neighboring rural environment.

There may be those who see in this situation only an ecologic relation, an 'urban–rural' conflict. Others who will see only a situation of contact between two cultures, between two ethnic groups with different economic resources, which would explain or even justify the pre-eminence of one ethnic group over the other. Yet this would be a mistaken view. The city's privileged position has its origin in the colonial period. It was founded by the conqueror to fulfill the very same function it still fulfils; to incorporate the Indian into the economy which the conqueror had brought and his descendants developed. The regional city was an instrument of conquest and is still an instrument of domination. It is not only a matter of 'contact' between two populations: the Indian and the Ladino are both integrated with a unique economic system, in a unique society.[44] It is for this reason that inter-ethnic relations, insofar as commercial activities are concerned, bear the characteristics of class relationships. The ecologic aspect of interaction between city and countryside, or between urban metropolis and community, in fact conceals specific social relationships between certain kinds of persons who hold differential positions with respect to the means of production and the distribution of wealth.

2.3 Social Stratification

There are essentially two ways in which to consider the relationships between Indians and Ladinos: that which only considers two ethnic groups, two cultures brought to a more or less close contact, which might be called the culturalistic perspective; and that which takes as its point of departure the existence of the whole society, of a single socioeconomic structure in which these two ethnic groups perform differentiated roles, and which might be called the structuralist perspective. The analysis made thus far is from the latter perspective. Yet this does not mean to deny the value of the culturalist approach. On the contrary, the perspective of cultural anthropology is valid when the analysis of social classes is set aside in order to consider other aspects of the relationships between the two ethnic groups.

In every society there may exist various systems of social stratification. Here it is possible to distinguish three systems of social stratification, that is, three social universes with respect to which social stratification may be studied: the Indian group, the Ladino group, and the total society in which Indians and Ladinos

[44] The word 'integration' is understood in its more general sense, that of being a functional part of a whole.

participate (that is, the inter-ethnic system). We may speak of two kinds of stratification: intra-ethnic and inter-ethnic.

2.3.1 *Intra-Ethnic Stratification*

Indians and Ladinos represent two different cultural communities. Each has a set of cultural values which may be called a value system. To the extent to which the value systems of these two communities are different, so too their systems of stratification shall likewise be different. It is thus easy to distinguish social stratification in each of them.

2.3.1.1 The Indians' Social Hierarchy

The Indian community is not stratified. All of its effective members equally participate in the same value system, and they are all equal with respect to each other. To participate in an effective manner in the Indian community means that Indians fulfill their duties in the community's political and religious structure.

The corporate community controls its members through control of its resources and through regular distribution of wealth. This is brought about through the cycle of religious festivities and through local government. Community government has traditionally been in the hands of principales, family and lineage chiefs who enjoy special prestige due to services rendered to the community, and sometimes due to special supernatural powers which are attributed to them by other members of the group.[45] The council of principales is a group of elders who enjoy an individual pre-eminence; it is not a social stratum. This form of government is linked to the original kinship organization, which is now disappearing. Its real power is decaying, and effective government is in the hands of the so-called Regional Council. This is the pinnacle of the double political-religious hierarchy (also called centripetal organization),[46] in which individuals climb to higher status by alternately holding civil and religious positions in the course of their lives. The individual named by his peers to hold a public position within this system is forced to accept it under the threat of strong social ostracism. Public functions imply a series of very heavy duties and monetary expenses. The selected individual (who always tries to escape from his functions before having been elected, but must rigorously submit to his duties once he has forcibly been sworn in) not only must abandon his farming, leaving it to the care of his family or even hired laborers, but must also spend large sums for festivities and ceremonies in the organization of which he

[45] G. Aguirre Beltrán, 1954: *Formas de gobierno indígena* (Mexico).

[46] F. Cámara Barbachano, 1952: "Religious and Political Organization", in: S. Tax (Ed.): *Heritage of Conquest* (Glencoe).

must participate. Passing through the hierarchy means years of indebtedness for many. When the public position is well performed it is a source of prestige and moral authority, but it does not bring major benefits. Personal power is strictly limited by the collectivity; authority is exercised for the benefit of the whole community and not for any restricted particular group.

It has been said that the expenses involved in festivities and ceremonies represent a prestige economy, that distribution of wealth (similar to Canadian potlatch and African bilaba) is the source of prestige.[47] Another author offers an opposite interpretation, which seems closer to reality: it is not wealth as such, but services rendered to the community which creates prestige, yet a certain amount of wealth is necessary to carry out these services adequately. Thus, there is not, strictly speaking, a prestige economy, since economic pre-eminence is not automatically translated into prestige. On the contrary, if a poor man performs his public functions well, he may achieve a status of great prestige in the community; that is if he finds the means to finance the festivities and ceremonies which are his charge, even when this may mean running into debts.[48]

Apparently economic pre-eminence of individuals is not favored by the community. We have seen that the means available to the Indian for accumulating capital are strictly limited. Also limited are the possibilities of investment. Basically, it is the corporate community itself which limits the economic possibilities of its members. In Chamula, members of the Council sometimes purposefully choose for the presidency individuals whose relative wealth is well known. This is obviously justified by the fact that wealthy persons can more easily perform their duties. But the social consequence of this act is the redistribution of wealth and maintenance of the 'principle of equality' in the group's social organization.[49]

Under these conditions it is impossible for a social stratum that stands out among the rest of the population to emerge in the traditional corporate community.[50] Individual economic pre-eminence is not transformed into prestige. It arises, individually, through positions held in the political-religious structure. The political organization of the community is a means to redistribute wealth and channel people's energy into service to the community.

It is important to qualify the phrase 'redistribution of wealth'. In effect, a fictitious redistribution occurs. It is nothing but elimination of likely economic pre-eminence of those individuals who for some reason have been able to accumulate a greater amount of goods than their peers. This wealth is not reabsorbed by the

---

[47] G. Aguirre Beltrán, 1954: *Formas de gobierno indígena, op. cit.*

[48] Ricardo Pozas, *Chamula, un pueblo indio de los Altos de Chiapas, op. cit.* In an interesting work recently published, F. Cancian proves that in Zinacantan (Mexico), the prestdige of a position depends on various factors which are difficult to measure among them the cost of the position, the authority it conveys, and 'idiosyncratic' factors. *Cf.* F. Cancian, 1963: "Informant error and Native Prestige Ranking in Zinacantan", in: *American Anthropologist*, 65,5.

[49] *Ibid.* Pozas attributes the principle of equality to vestiges of clan organization.

[50] Cancian (*loc. cit.*) suggests that in Zinacantan there does exist a rudimentary 'economic stratification'.

community. It is consumed in liquor, ceremonial clothing, fire-crackers and fireworks, and in hundreds of articles employed in what an observer has named 'institutionalized waste'.[51] These expenses required by the ceremonial economy associated with the functioning of the political and religious organization are transformed into income for those who provide these articles for the community. These purveyors are urbanized Ladinos, many of whom are craftsmen specialized in the kinds of articles consumed by Indians. Aguirre Beltran even states that trading of these ceremonial articles is, in Chiapas, "the real source of life of a city of 18 thousand inhabitants".[52] We may thus conclude that the structure which maintains equality within the Indian community, preventing the emergence of social classes, also contributes to the whole Indian community's dependence on the city, that is, to the differentiation of social classes between Indians and Ladinos.

There exists in the region yet another form of government: the Constitutional Council, which is a part of the national political regime and the only 'legal' government, from the point of view of the national constitution. This is the link which unites the community to other political institutions such as political parties, regional and national legislatures and national executive power. It is the means employed by national governments to extend their administrative and political control over Indian populations.

The constitutional council is generally controlled by Ladinos, even though the municipal president may be an Indian. Local Indian government will surely disappear in time, to be substituted for by the Constitutional Council. To the extent to which the Indians participate more and more in national politics and in official governmental organisms, the Constitutional Council is likely to become a means of social differentiation within the Indian community, perhaps creating a higher stratum of 'court clerks' and functionaries.[53]

2.3.1.2 Social Strata Among Ladinos

Ladino society, as every 'Western' society, is stratified. This stratification is influenced by such factors as land ownership, income, occupation, education, and family lineage. The Ladino city is highly differentiated in terms of these diverse criteria, even having its own local aristocracy descending (in fact or in fiction) from important colonial families. Status indices are correlated with one another.

---

51 G. Aguirre Beltrán, 1954: *Formas de gobierno indígena, op. cit.*, p. 103.

52 *Ibid.*

53 In Chiapas, the Instítuto Nacional Indígenista de México is training young Indians as municipal secretaries for the positions held by the Ladinos. In Guatemala, the penetration of the national political parties into the Indian communities during the democratic regimes of the 1944–54 decade modified the traditional structure. These problems have been treated in a collective work which the author was unfortunately unable to consult while working on this essay: *Political changes in Guatemalan Indian communities* (New Orleans, 1957).

The family line, large land ownership, big business, and participation in local politics go together. But on the other hand, a high level of education (especially university) is more typical of the 'new rich', the professionals (physicians, lawyers, engineers), who are new to the region but are developing other more traditional interests, and thus frequently associate with the older families through marriage.

It would be arbitrary to determine the number of strata existing in the Ladino society. In Guatemala, Adams indicates five 'primary economic types': large landowner, plantation owner, medium-sized landowner, small landowner, tenant, and worker. The last three levels often overlap and may be treated as a single stratum. Workers are in turn divided into tenant farmers and day laborers. But on the other hand the same author also speaks of only four strata: the upper cosmopolitan, upper, middle and lower 'classes'.[54] In Jilotepeque, Tumin differentiates three strata, according to wealth, family prestige, and other characteristics. Combining indices of various scales, which he then divides into three groups: an upper 'class', with 45.5 % of the Ladinos in his sample; a middle 'class', with 40.9 %, and a lower 'class', with 13.6 %. Applying the same indices to a sample of Indian population, the result is no upper class at all, and the concentration of two-thirds of the sample in the 'lower class'. Nonetheless, on Tumin's scale a certain number of Indians and Ladinos hold identical positions.[55]

In terms of our stratification analysis, this exercise in status classification is of only limited value. We have already seen that the Indian community is socially unstratified and Tumin confirms it in his analysis. Tumin's statistical exercise is useful only to establish 'standards of living' which may have no major social implications (such is the case, in effect, among the Indians). And with respect to Ladinos, Tumin admits the weakness of his own analysis by showing that in Jilotepeque, Ladinos are in fact divided into only two strata perceived by everyone: the elite, called society, composed of 20 families (less than 20 % of the Ladino population), and the populace. At the lowest level of the Ladino ethnic group, it is difficult to distinguish clearly a Ladino from an Indian. In Panajachel, Tax also speaks of two Ladino classes: the 'upper urban bourgeoisie' and the 'lower rural'.[56] In other communities there also exist specifically defined strata.

Ladinos place high value on wealth and property, which are one of their raisons d'être. These values constitute the foundation of all of their economic activity. Ladino society is mobile, and opportunities for upward mobility exist, in principle, for everyone. As opposed to the Indian, the Ladino conceives his own society as a stratified system. Certain activities, especially manual occupations, belong to an inferior order and must be avoided; there are others, especially commerce, to which they aspire. Finally, the condition of landowner is the most envied. The

---

[54] Richard N. Adams, 1956: *Encuesta sobre la cultura de los Ladinos en Guatemala* (Guatemala: EMEP, 1956).

[55] Melvin Tumin, *Caste in Peasant Society, op. cit.*

[56] Sol Tax, *Penny Capitalism, op. cit.*

'good family' plays an important part in these provincial societies, and the fact of being related, through kinship, marriage or *compradago*, to important families is obviously a way of acquiring a high social status. Ladino culture, as opposed to the Indian, is highly competitive and authoritarian.[57]

2.3.2 Inter-Ethnic Stratification

Stratification means that certain characteristics or variables are unequally distributed among individuals. The combination of some of these characteristics and the value attributed to them by members of society account for the existence of a scale or continuum, in which individuals occupy higher or lower positions with respect to one another. If a set of individuals have in common a set of these characteristics, which distinguish them from other groupings, and if this is recognized as such by society, we may then speak of a stratum or social class. When a stratified system has quantifiable status characteristics, and is homogeneous from a cultural and racial point of view, some authors commonly refer to it as a 'social class system'. But if other factors are involved, and if the status indices are associated with qualitative factors such as 'race' or culture, then some specialists speak of a 'caste system'.

Ladinos and Indians hold different positions in the stratification scale, according to such well known variables as income, property, degree of education, standard of living, etc. Given the fact that Ladinos concentrate along the scale's upper ranks and Indians along the lower ones, the two ethnic groups may be considered as strata within one stratified system. They are in effect the only strata in this system, because in the value systems of both groups ethnic characteristics (cultural and sometimes even biological) play a more important part in stratification than do other criteria. Ladinos hold a higher position not only in the objective scale of socioeconomic characteristics, but they also consider themselves, qua Ladinos, as being superior to the Indians. They are contemptuous of the Indian as such. The latter, on the other hand, are conscious of their social and economic inferiority. They know that those traits which identify them as Indians place them in a position of inferiority with respect to Ladinos.

Even while stratification is objectively presented as a scale or continuum, it in fact functions socially as a system with only two strata which are characterized in cultural and biological terms. Ladinos make use of physical stereotypes to affirm their 'whiteness' in contrast to the darker Indians. As Tumin has pointed out, it is a matter of ideal types, since the Ladino population is in effect a *mestizo* one. This fact notwithstanding, one of the most valued criteria among the higher Ladino strata is that of their supposed 'Spanish blood'. Other observers have noted that, in

[57] B. Colby and Van den Berghe, 1961: "Ethnic relations in the Southeastern Mexico", in: *American Anthropologist*, 53.4.

San Cristobal de las Casas, there appears to be a coincidence between the socio-economic scale and the biological continuum.[58] Racial criteria, nonetheless, do not perform an important role, precisely because it is impossible to classify the population in either ethnic group on an exclusively physical basis. Cultural factors are essential to stratification: in the first place comes language and dress, but there is also self-identification and personal identification by others. Thus, mastery of Spanish and changes in dress do not ipso facto turn the Indian into a Ladino. Essentially the Indian condition lies in his being integrated to his Indian (corporate) community, and participating in the traditional social structure (kinship groups, civic-religious hierarchy). It is the 'cultural' and not the 'biological' Indian who constitutes the lowest stratum. The Indian is conscious of this situation. Learning Spanish not only represents for him a means of upward mobility, but also an instrument of defense in his daily relationships with Ladinos. The adoption of Ladino dress styles also reduces the stigma of his inferior condition in his relationship with Ladinos. (Let us disregard here a discussion of psychological counter-acculturation, represented by a sharp rejection of everything which is Ladino, a phenomenon which often appears among the more conservative elements of the Indian community.)

The definition of the two ethnic groups depends upon strictly cultural factors which, due to their historical importance in the region, subsume and impose themselves upon all other factors of stratification. While it dichotomizes social relationships, ethnic stratification diminishes the importance of the socio-economic scale or continuum based on quantitative indices. To such a degree that many Indians and Ladinos share the same socio-economic level without the disappearance of ethnic stratification. Robert Redfield noted that in a Guatemalan village, "the greater the Ladinos' upward mobility, the more they tended to be contemptuous of the Indians and to identify lower-class Ladinos with Indians".[59] And, naturally, those 'lower-class' Ladinos considered themselves superior to Indians.

These cultural values are reflected in inter-ethnic relations. Ladinos always behave in an authoritarian or paternalistic manner towards Indians. These are treated with familiarity, yet it is expected of them to show signs of respect and submission. Unskilled manual labor is considered an attribute of the Indian. Notwithstanding legal equality proclaimed in the Constitution, Indians are subject to discrimination, particularly in the cities, where they are exposed to all kinds of arbitrary and humiliating behavior by the Ladino population.

Effective social contacts between Indians and Ladinos are, with the exception of the already mentioned economic relations, very limited. There exists no real social interaction between the two ethnic groups. Traditional religious and political activities are performed separately; common participation at parties and sports is

---

58 B. Colby and Van den Berghe, *loc. cit.*

59 Robert Redfield, 1956: "The Relations Between Indians and Ladinos in Aqua Escondida, Guatemala", in: *América Indígena*, 16,4.

almost nonexistent. The only non-economic relationship in which Indians and Ladinos formally participate is *compadrazgo*, yet as has already been pointed out, here too the Indian's inferiority is obvious, and here too there are economic implications.

2.4 Social Mobility

There is upward mobility from the Indian stratum to the Ladino; but its nature and characteristics are by no means simple and they vary from region to region. A public opinion poll carried out by Tumin in Jilotepeque showed that there are relatively more Indians than Ladinos who believe that movement from one group to the other is possible. Indians tend to believe they can achieve this through the accumulation of wealth, while Ladinos believe that the modification of strictly cultural characteristics is needed. Given the Ladinos' superiority, they have an interest in checking the Indians' mobility. Adams has pointed out that in a community where cultural differences between Indians and Ladinos are small, the latter resort to a whole series of ruses in order to maintain their superiority—even the invocation of 'racial' factors where no biological differences exist.

Upward mobility among Indians represents a process of acculturation.[60] But learning Spanish and adopting Ladino dress styles is insufficient. The Indian must also become socially (generally meaning physically) separated from his community. In order to become a Ladino, the mobile Indian must cut his ties with the social structure of his corporate community. He must not only modify his cultural characteristics, but also his 'social' condition as an Indian. It is very unlikely and one might say impossible, for an Indian to become a Ladino in the midst of his own community. The 'ladinized' Indian is a marginal man. Well known are cases of Indians in the process of acculturation, who wear Ladino clothes when going to the city, and change again into their Indian costume upon returning to their community. The difficulties encountered by the cultural promoters of the I*nstituto Nacional Indigenista* in Mexico are also well known. It should be noted that these promoters, in their positions as teachers, nurses, and practical farmers at the service of the State, come to achieve a higher socio-economic status than the local Ladinos. This suggests that mobility increases when the community's traditional structure begins to disintegrate. Researchers have pointed to the existence of diverse stages in the Indian's acculturation process. We thus speak of the traditional Indian, the modified Indian, the ladinized Indian, the sheathed Indian, etc. These are descriptive categories rather than analytical ones, and since they possess such diverse connotations, they should be handled with great care. On the other

[60] We use the terms 'transculturation' and 'acculturation' interchangeably, in the sense in which the latter is used by g. Aguirre Beltrán, 1957, in: *El Proceso de Aculturación* (Mexico).

hand, there are also 'indianized' Ladinos and, to be sure, Ladino culture as such contains innumerable cultural elements of Indian origin.

The Indian's upward mobility means both a process of acculturation and an elevation in the socio-economic scale. It is neither the poorer Indians nor the subsistence farmers who become.

2.4.1 Ladinos

To become a Ladino in a cultural sense also means being a trader or regularly producing for the market and, in general, acquiring a higher standard of living. This does not mean that all of those who become traders or sell their produce in the market or who achieve a better standard of living necessarily become Ladinos. Nor does it mean that Ladinos who descend the socio-economic scale become Indians. In effect, a Ladino will always be a Ladino, low as he may fall in the socio-economic scale. But an Indian, provided that he ascend the socio-economic scale, may become a Ladino; what is more, he will never be a Ladino unless he ascends on the socio-economic scale (that is, unless he obtains higher indices on the objective hierarchies of social status). Hypothetically the Indians may ascend the socio-economic scale without becoming Ladinos. This occurs in the case of a general rise in the community's prosperity, provided that it maintains its Indian cultural characteristics. This situation could be the result of community development programs, but only if the directors at the same time applied a deliberate policy of conserving and stimulating the Indian culture. This is not the case at present.

According to the perspective which is adopted, inter-ethnic stratification may be considered as a scale (composed of various levels), as a continuum (a series of quantitatively different positions), or as a dichotomy. In social life these perspectives cut across each other. For the Indian moving upward within the stratification system, inter-ethnic mobility represents both a gradual or quantitative evolution (his income increases, he improves his house, he buys a pair of shoes, he learns how to read and write in Spanish, etc.) and a radical metamorphosis, a qualitative 'leap' (he abandons his community, earns a salary in the city, marries a Ladino woman, denies his origins). At which point of the individual's cultural evolution does this metamorphosis take place? It varies according to circumstances. It is obvious that when the mobile Indian's point of departure lies high in the socio-economic scale, ethnic transformation will occur with a certain smoothness. The individual departing from a lower level, on the other hand, may accelerate the process by breaking off with his community and migrating to another region. But in this case he is placed outside of the given stratification system, and thus his transformation is not, strictly speaking, upward mobility within a given system of social stratification. Frequency and speed of mobility also depend on other factors: rigidity of the community's traditional structure, rigidity of the ethnic barrier maintained by Ladinos, the region's economic situation, and finally, the effectiveness of Indianist policy.

2.5 The Dynamics of Inter-Ethnic Relations: Classes, Colonialism, and Acculturation

Let us pull together the different threads in this essay and attempt a general formulation of the system of relationships between Indians and Ladinos. Our historical point of departure will be the Spanish Conquest, although we do not deny the importance of pre-Hispanic social processes in the subsequent character of the Mayan region. The Spanish Conquest was a military enterprise and part of the political and economic expansion in post-feudal and mercantilistic Europe. The Conquest was fundamentally influenced by commercial factors (the lust for gold and spice). As a military enterprise the Spanish Conquest was a violent confrontation of two societies, two different cultures. The weaker one—the Indian—succumbed. The Indians received from the conqueror the treatment accorded since ancient times to the vanquished: looting, dispossession, slavery, even extermination. Yet the Conquest of the New World was not like preceding ones. In Spain, deep transformations were taking place due to the Reconquista. The American continent would perform an essential role in Europe's economic development, and to the native populations were ascribed specific functions in this development. For different political and economic reasons, destruction and enslavement of native populations had to comp to a stop. The military conquest was transformed into a colonial system. Just as other colonial systems which the world has known since then, this one was managed over three centuries on behalf of the interests of certain powerful social classes of the metropolis, and that of their representatives in New Spain. The Crown's policy reflected these changing and often conflicting interests.

At first Indian chiefs and Indian aristocracy were kept in their positions, which suited the colonial administration's realpolitik. But towards the end of the sixteenth century Indian communities had become socially and economically homogeneous. Their internal social differentiation was no longer in the interests of the colonizer. Residential segregation of Indians (through settlements of converted Indians and other mechanisms) and the encomiendas (lands which the Crown granted as trusteeship to the conquistadores) were the first instruments used by the conquistador to levy taxes and services. Part of the Indian society's wealth was simply transferred to the conquering society. Indian communities were transformed into labor reserves of the colonial economy. Systems of serfdom and forced labor in plantations, mines, and workshop constituted the basis of the economic system.

Colonial society was the product of mercantilist expansion: of the dawning of the bourgeois revolution in Europe. Its structure still retained much of the feudal era, especially in the character of human relationships. Some researchers even affirm that feudalism grew stronger in America after it had begun to decline in Spain, and that America 'feudalized' Spain once again.[61] Exploitation of the Indian population constituted one of the main goals of colonial economic policy. In order to maintain this labor reserve, it was framed by a complex of laws, norms, restrictions, and prohibitions which kept accumulating during three centuries of colonialism, and which resulted in the corporate 'folk' communities. All things were determined for the settler's benefit: the land tenure of the Indian community, its local government, technology, economic production, commerce, residential pattern, marriage norms, education, dress styles, and even its idiom and use of language. In Spain, nobles, landowners, commercial bourgeoisie and petty bourgeoisie were at times fighting, at times co-operating in the struggle for their respective interests. But in Spanish America a rigid social hierarchy based upon centralization of political and economic power and validated in the Legislation of Indians kept the natives in their position of inferiority with respect to all of the other social levels.

The colonial system worked on two levels. The restrictions and economic prohibitions which Spain imposed upon her colonies (and which were to foment the Independence movements) were repeated, often aggravated, in the relations between the colonial society and the Indian communities. The same commercial monopolies, the same restrictions on production, the same political controls which Spain exerted upon the Colony, the colonists imposed upon Indian communities. As Spain was to the Colony, so the Colony was to Indian communities: a colonial metropolis. Since then mercantilism penetrated even the most isolated villages of Spanish America.

The social groups in Spanish America which took part in the processes of economic production and distribution which sustained the Spanish Empire, also participated in the class structure of the colonial system. In the same way the Indian population participated in the class structure of the Colony. Colonial relationships and class relationships underlay ethnic relationships. In terms of colonial relationships, the Indian society as a whole confronted colonial society. Primary characteristics of the colonial situation were ethnic discrimination, political dependence, social inferiority, residential segregation, economic subjection, and juridical incapacity. In the same way, class structure was defined in terms of labor and property relations. These relations were not defined in ethnic, political, social, or residential terms. Only juridical coercion (supported by military power) as well as other economic and extra-economic pressures intervened in the establishment of labor relations. Labor relations were not between two societies, but only between two specific sectors within them. Colonial and class relationships appear intermixed throughout this period. While the former primarily answered to mercantilist interests, the latter met the capitalist ones. Both kinds of relationships were also opposed to each other: the development of class relationships came into

[61] Angel Palerm, 1952: "Notas sobre la clase media en México", in: *Ciencias Sociales* (Washington); no. 14–15 and 16–17 (Reproduced in: *Las clases sociales en Mexico*, s.f. (1960)).

conflict with the maintenance of colonial relationships. Indian communities were constantly losing members to the developing national society. Despite tutelary legislation, the biologic and cultural mixing was a constant process which kept producing new problems for colonial society. Those Indians who for various reasons were absorbed by the larger society, therefore, quit the aforementioned colonial relationships to become integrated simply in a class structure. In consequence, they were no longer Indians.

These two kinds of socio-economic relationships in which the Indian ethnic groups were involved received moral sanction with the rigid social stratification in which the Indian (biologically, culturally, and juridically defined) was always at the bottom (with the exception of the slave). From these conditions there emerged the corporate community and the formation of indo-colonial cultural characteristics, which we today call Indian culture. Ethnic relationships of the period thus presented three main aspects: two kinds of relationships of dependence and one kind of relationship of order.[62]

The dynamics of these systems of relationships were varied. The colonial relationships between Indian communities and the larger society tended to strengthen the Indian communities and foment their ethnic identity. The subordinate group usually reacts to a dominant-subordinate relationship of the colonial kind with a struggle for liberation (at the most diverse levels). Colonialism produces nationalism and struggles for independence. The colonial period was not devoid of native rebellions. Conversely, class relations contributed to the disintegration of the Indian community and its integration to the larger society. Both kinds of relations complemented each other in terms of the Indian's oppression. But the opposed tendencies which they engendered explain why certain Indian communities survived, while others were transformed into peons' or squatters' enclaves, in the haciendas which displaced the encomiendas of the sixteenth and seventeenth centuries. Colonial relationships usually dominated class relationships. Although colonial relations were only one aspect of a world-wide system of mercantilist class relations, the more narrowly defined class relationships between Indians and Spaniards (including *criollos*, Spaniards born in the Colony) usually appeared in the form of the colonial relations described above. This was essentially due to the nature of colonial economy.

Finally, social stratification, which has sometimes, because of its rigidity, been called a caste system, reflected more the colonial character than the class character of the Indian's subjugation. The stratification system, in turn, exerted its own influence upon the development of class relationships.

Political independence in Spanish America did not basically change the relationships between Indians and the larger society. Despite the legal equality of all citizens (including Indians), various factors joined to maintain the 'colonial'

---

[62] On the concepts of relation of dependence and relation of order and their application to the study of class structures, see S. Ossowski, 1963: *Class Structure and Social Consciousness* (London).

character of these relations. First, internal struggles which lasted many decades and second, the economic depression during the first half of the nineteenth century. Both kinds of factors helped to keep Indian communities marginal, isolated from the outside world, and increasingly corporate. Another reason should also be taken into account. At the beginning of the colonial period tutelary laws were established because it was considered that Indians were inferior beings. But by the end of three centuries of colonialism, these laws had served to maintain and fix that inferiority. In consequence, when legal equality was declared, the Indian was effectively in a condition of inferiority to the rest of the population, in every area of economic and social life.

The first effective changes occurred during the second half of the nineteenth century: first with the Reform laws and later with the introduction of new cash crops (principally coffee) into the Indian region. Both phenomena, of course, are closely related to one another. Legal equality and disamortization of communal land had two immediate consequences: the Indian could now freely dispose of himself in the labor market, and the land he held could become private property. In fact, this did not take place in the abstract, but in the specific situations that have already been mentioned: extension of commercial farming; penetration by Ladinos into communities inhabited by Indian ethnic groups; appropriation of land by Ladinos; formation of great latifundia and the Indians' wage labor on these properties and haciendas. Coffee plantations became working centers for a considerable mass of Indians, legally or illegally recruited from their communities. At the same time the first products of industrialization penetrated into the more distant villages of the Indian region in the form of goods carried by Ladino traders. In this way new economic relationships were established between the Indians and the rest of the population.

Expansion of the capitalist economy during the second half of the nineteenth century, together with the ideology of economic liberalism, once again transformed the quality of ethnic relationships between Indians and Ladinos. We consider this stage as a second form of colonialism, which we might call internal colonialism. Indians of traditional communities found themselves once again in the role of a colonized people: they lost their lands, were forced to work for the 'strangers', were integrated against their will to a new monetary economy, and fell under new forms of political domination. This time, colonial society was national society itself, which progressively extended its control over its own territory.[63] Now there were not only isolated Indians who, abandoning their communities, joined the national society; but Indian communities themselves, as a group, were progressively incorporated to expanding regional economic systems. To the extent to which national society extended its control, and capitalist economy dominated

[63] Pablo González Casanova, in a different and independent analysis, also brings forth the existence of internal colonialism in Mexico. The present essay bears a particular case, which may be considered within González Casanova's general approach. See his study, "Internal Colonialism and National Development", in: *Studies in Comparative International Development*, 1,4 (1965).

the area, relations between colonizer and colonized, between Ladino and Indian, were transformed into class relationships.

The corporate community has been characteristic of colonial society in Indian America. Corporative social structure has an ecologic and economic basis. When colonial society is transformed into 'underdeveloped' society, when the economic structure of the corporate community is modified (loss of lands, wage labor, commercialization of agricultural produce, etc.), then it is rather unlikely that the corporate quality of the community's internal social relationships should survive for long. As we have seen, some of the Indian's cultural characteristics are bound to the highly structured corporate community. If this structure should progressively disappear, these cultural characteristics would become weaker.

Ethnic stratification in the region is the result of this historical evolution. It reflects the colonial situation which has been maintained till present times. Behind inter-ethnic relationships, which show themselves as a stratification system, there is a social class structure. When an Indian works for a Ladino, the main point is not the inter-ethnic relationship but the labor relation. During the decade of the thirties, the Indians of Chiapas organized to defend their working conditions in the coffee plantations; not as Indians, but as workers. During the years 1944–1954 there were also labor unions of Indian agricultural workers in Guatemala. They have become organized in their struggle for land, under the agrarian reform programs but as landless peasants. These relationships sometimes assume cultural shapes. The struggle for land, for instance, is carried on in the name of restitution of communal and clan lands. At times there have also emerged messianic movements against Ladinos. Yet it was always a matter of structural changes within the traditional community.

Inter-ethnic stratification no longer completely corresponds to new class relationships which have developed along with a monetary economy. 'Colonized' Indians are not a social class. We are not saying that Indians and Ladinos are simply two social classes. This would be over-simplifying a deeply complex historical situation. During the course of economic development (or more precisely, of the development of economic underdevelopment, as a result of colonial economy), various new social classes emerge. They are not yet totally formed, because 'colonial' relationships still determine the social structure at different levels. The Indian participates in various kinds of socio-economic relationships. He holds various occupational roles at the same time. He may be a small farmer in the communal lands, an ambulant trader, a salaried worker during different periods of the year, or during the course of his life. This situation may last as long as the regional economic structure allows it. But this structure is suffering rapid changes: monetary economy is expanding, capitalist labor and trade relations are becoming generalized, regional communications are developing, and local industrialization is getting started. These different kinds of class relationships contribute to separate the individual from his corporate community. The community's corporate structure is breaking up. Should it disappear, inter-ethnic stratification will have lost its objective basis.

Nonetheless, the inter-ethnic stratification system which, like every stratification system, is deeply rooted in the values held by the members of the society, is an essentially conservative force within the social structure. While it reflects a situation of the past (the clear dichotomy between Indians and Ladinos in every area of social, economic, and political life, characteristic of the colonial situation), it curbs the development of new class relationships. We should not forget that the landless peasant and the salaried worker are also Indians. Even though relations of production will be determinant of future transformations in the region, ethnic consciousness may weigh heavier than class consciousness. Thus, exploited or poor as a Ladino may be, he feels privileged as compared to the Indians, even those who may have a standard of living higher than his own. Indians, on the other hand, tend to attribute all of their misfortunes to the Ladinos as such (a position which, by the way, is shared by certain romantic indigenous intellectuals), an attitude which contributes to the concealment of objective relationships between classes. This range of problems has been little studied in the region and it represents, in my opinion, an interesting field of research.

To the extent to which class relationships become more clearly defined, there emerges a new stratification, based on socio-economic indices. This stratification already exists among Ladinos, and is progressively expanding to the Indians. The status symbols of the Ladinos are beginning to be valued by the Indians too. It is no longer sufficient—or even desirable—that the Indian should become 'ladinized'. Young Indians, particularly those who now work for the Government, without ever breaking ties with their communities, are buying dark sunglasses, pens, watches, etc., wearing them ostentatiously as symbols of prestige. The situation will have radically changed when social stratification includes Ladinos and Indians independent of their ethnic characteristics. Ideally this would mean the maintenance of Indian cultural identity independent of stratification. To what degree this situation is workable depends on many special factors. It has been noted that in Quetzaltenango (Guatemala) something of the sort is taking place, and this also seems to be the case in Mexico among the Maya of Yucatan, the Zapotec of Oaxaca, and the Tarascans of Michoacän.

This also depends on the attitudes and reactions of Ladinos, whose position is not stable within the class society. Ladinos have always accepted (at least from one generation to the other) the admission of acculturated Indians into their group. It is difficult to foresee reactions of the Ladino community faced with two hypothetical alternatives of the interethnic stratification system's evolution: on the one hand, the complete assimilation of Indians (which is rather unlikely); and on the other, a general economic rise of the Indian ethnic group as such (which would be a challenge to Ladino superiority). Development of a class society leads toward either of these hypothetic situations. The final result will depend on how class conflicts are solved. Indian- Ladino acculturation is a process operating on several levels. Adams foresees the ladinization of Guatemala, while in Mexico there is some talk about the integration of Indians into the Ladino culture. Yet it is necessary to study which aspects of Indian culture will be transformed in this process. Here it is convenient to distinguish structural from cultural. Those cultural elements intimately associated with the corporate structure of the community and with inter-ethnic stratification will surely disappear with the transformation of the colonial situation into a class situation. In this sense, the Indian will stop being an Indian (or will only be so in a cultural sense, and no longer in a social or structural sense). Tax has pointed out that in Guatemala social relations are 'civilized', while the world view remains 'primitive'.[64]

There may also exist a class culture, and many 'Indian' cultural elements will accompany the development of class society as elements integrated to a new structure. One author has recently suggested that the 'Indian' culture of Chiapas is nothing but a 'rural' culture, similar to rural cultures in other parts of the world.[65]

The system of inter-ethnic stratification can only be understood as referred to the corporate structure of the Indian community and its cultural characteristics. This structure, in turn, can only be explained in terms of its colonial past. The colonial situation has become progressively transformed. The Indian thus finds himself in the midst of diverse and contradicting situations: at times he is 'colonized', and at times he is a member of a class (in the sense that he is in a typical class situation).[66] In other words, not only does the Indian perform various roles (as everybody else), but he also participates in dichotomized role systems, which are historically and structurally conflicting.[67] Nor does the Ladino escape ambiguity: at times he is the 'colonizer', at times bearer of 'national culture' and member of 'national society', and at the same time he finds himself in most diverse Class situations, in confrontation with Indians and other Ladinos.

Until now our analysis has mainly focused on corporate community as prototype of one of the poles of inter-ethnic relationships. This position is obviously inadequate. It overlooks, at the cultural level of inter-ethnic relationships, those 'cultural' Indians who are not incorporated into a corporate community; that is, those 'modified', 'Latinized', 'acculturated' categories referred to by the anthropologists. Nonetheless, this approach finds its justifications from the fact that the analysis was not carried out on the cultural but rather on the structural level. On the other hand, it has been stressed that two main structural units are involved in

---

64 Sol Tax, 1956: "La Visión del mundo y las relaciones sociales en Guatemala", in: *Cultura Indigena de Guatemala* (Guatemala: EMEP).

65 V. Goldkind. 1963: "Ethnic Relations in Southeastern Mexico: A Methdological Note", in: *American Anthropologist*, 65,2.

66 We use the term 'class situation' not in the sense given by Max Weber (*Cf.* H.H. Gerth and C. W. Mills (Eds.), 1946: *From Max Weber. Essays in Sociology* (New York: Oxford University Press): 181, but in the sense that the individual who finds himself in such a situation participates with others in a kind of relations having the character of class relations.

67 See S.F. Nadel, 1957: *The Theory of Social Structure* (London), especially chapter IV. It would be interesting to do a formal analysis of the roles of inter-ethnic situation here described. Nadel's model, nonetheless, does not seem to include a situation as that which is brought about between Indians and Ladinos when they face each other as colonizer and colonized and as belonging to opposite classes *simultaneously*. In there words, the same process of interaction between individuals andgroups may be understood at different levels of an analysis of roles and in varying conceptual terms. Nadel's concept of 'summation' comes closest to this situation.

the structure of inter-ethnic relations: the corporate community and society as a whole (in its diverse manifestations). The task now remains to approach the problem from the point of view of the total society.

Contemporary inter-ethnic relations partly result from colonial policy. They also represent the disintegration of that policy and are a function of present economic and class structures. As has been shown by various economists, underdeveloped economies tend to polarize into areas of growth and structurally related areas of stagnation. The Maya region of Chiapas and Guatemala constitutes such an area, as do other Indian areas of Mexico. The 'marginal' populations inhabiting these areas are growing in absolute numbers, despite national economic development.[68] If this happens in Mexico, despite accelerated economic growth in recent years, then in Guatemala, where there has been no such development, it must surely happen with greater intensity. During the colonial period, colonial relations in the Indian regions served the interests of a well defined dominant class which in turn subdued the colonial society as a whole to its own interests, insofar as relations with Spain would permit. In the situation of internal colonialism (which might be called the endo-colonial situation) class relationships within the whole society are more complex. The regional dominant class, represented by Ladinos, is not necessarily the dominant one in the national society. In Guatemala, since the defeat of the nationalist bourgeoisie in 1954, these two groups became identified. There is no contradiction between landowners, commercial bourgeoisie (particularly coffee growers) and foreign capital.[69] Eighty nine In Mexico the situation is different. National power is held by a bureaucratic, 'developmentist' bourgeoisie, a product of the 1910 Revolution. This bourgeoisie has displaced latifundists on a national level, but in more backward regions, such as Chiapas, it tolerates them while seeking the support of a new rural bourgeoisie composed of traders, neo-latifundists and public employees.[70] In both Mexico and Guatemala the regional dominant class is composed of 'power brokers'—to use Wolf's term[71]—of mestizo origin who have come to fill the power vacuum left by the old feudal landowning aristocracy. In Guatemala the endo-colonial situation is stronger than in Mexico, where latent contradictions between the 'developmentist' bourgeoisie in power and its weak shadow in the Indian hinterland contribute to a rapid development of class relationships to the detriment of colonial relationships, and have allowed the development of a structural development-under- development dichotomy. Thus, inter-ethnic relations at the level of total society may be considered as a function of

[68] *Cf.* Pablo González Casanova, 1962: "Sociedad plural y desarrollo: el caso de México", in: *América Latina*, 5,4.

[69] Jaime Diaz Rozzotto, 1958: *El character de la revolución guatemalteca* (Mexico). Also see Richard N. Adams, 1960: "Social Change in Guatemala and U.S. Policy", in: *Social Change in Latin America Today* (New York).

[70] *Cf.* Rodolfo Stavenhagen, 1963: "La réforme agraire et les classes rurales au Mexique", in: *Cahiers Internationaux de Sociologie*, 34.

[71] Eric Wolf, 1959: *Sons of the Shaking Earth* (Chicago).

the development-underdevelopment structural dichotomy (in its social aspect of internal colonialism), and of the dynamics of national class structure.

For purposes of analysis, four elements may be isolated in the inter-ethnic situation: colonial relationships, class relationships, social stratification, and the acculturation process. These four elements constitute interdependent variables and with them we may attempt to build a hypothetic model of interethnic relations.

2.5.1 Colonial Relationships

These relationships are a function of the structural development-underdevelopment dichotomy and they tend to be in force for as long as the dichotomy persists. As long as there are areas performing as internal colonies in underdeveloped countries, the relationships characterizing their inhabitants tend to take the form of colonial relationships. These are strengthened where there exist, as in the Maya region, marked cultural differences between two sectors of the population, leading to a rigid stratification defined in cultural and biologic terms (which is sometimes called caste). Colonial relations tend to limit and impede acculturation, cultural ladinization, and to maintain a rigid stratification. There exists an obvious interest on the part of the dominant ethnic group (Ladinos) in maintaining colonial relations, especially when their predominance depends on the existence of cheap and abundant labor. This is the case when possibilities of expansion of the economy are few, when agriculture has a low level of productivity and when the labor-capital relation in agriculture is high, when local or regional industrialization is weak or nonexistent; and when the region's internal market is poorly developed. Therefore the maintenance of colonial relations is rather a function of the degree of development of national economy than of local or regional decisions.

In contrast to Ladinos, the Indians—the subordinate ethnic group—derive no benefit from the colonial situation and may try various forms of reaction to it. The first is withdrawal into the corporate community, both physically and socially. As Wolf pointed out, this has happened on various occasions in the history of the region, and it represents on the part of the Indian ethnic group a latent tendency which becomes manifest when the economic and political situation allows it. In association with this withdrawal, the Indians also react to the colonial situation in terms of 'nationalism'. This form of reaction may have as its objective the strengthening of the Indian government (regional council), and possibly the struggle for the Indians' national political representation. It also becomes manifest through measures adopted to encourage education in the Indian language and development of Indian culture. It particularly becomes manifest through an extreme anti-ladinism and resistance to ladinization. Here there also intervene other counter- acculturative factors such as messianism and, on certain occasions, armed upheavals and other violent manifestations. Finally, there is a third form of reaction to the colonial situation, and this is assimilation. It is an individual process which, as has been seen, represents a separation from the corporate structure of the community. From a cultural point of view it represents ladinization. From a structural point of view it means that the individual becomes integrated to the class structure, no longer as an Indian (that is, a colonized person), but simply due to his relationship to the means of production. Ladinization, as we have seen, may be the result of upward mobility in the scale of socio-economic indices. But generally it only means the proletarianization of the Indian.

Of the three main forms of reaction to the colonial situation, the first, simple withdrawal, does not seem to have many adherents at present. Among those who are still clinging to it we find a few traditionalistic elders. But other members of the community know that there are better ways to combat the harmful effects which colonial relations have upon Indians. The reaction which we have called 'nationalism' (for lack of a better term) assumes diverse shape? Some of them are spontaneous and circumstantial (such as armed upheavals and messianic movements); others have been induced by external agents (such as education in the Indian language); and still others may be the consequence of a political consciousness of Indian communities (such as the election of a person participating in corporate civic-religious political structure, to a position in the constitutional municipal government). At present, the main forms of 'nationalistic' reaction are promoted—at least in Mexico—by the national government's specialized agencies. Measures such as literacy in the Indian language and adequate political representation of the Indians show that those responsible for *indianist* policy are conscious of the colonial character of inter-ethnic relations, despite the fact that the problem has never been formulated in those terms by the ideologists of indigenismo. Yet paradoxically, these measures are only taken as a means to an end which represents its absolute negation, that is, the incorporation of the Indian to Mexican nationality, in other words, the disappearance of the Indian as such. The paradox, nonetheless, has a practical justification: national integration can only be achieved if contradictions inherent to colonial relations are overcome. This can be done either by suppressing one of the terms of the contradiction, or by a qualitative change of content in that relation. By encouraging measures of a 'nationalistic' kind, indianist policy is committed to the second of these alternatives. Yet if the contradiction inherent to the colonial relation between Indians and Ladinos is solved, there would be a greater contradiction solved at the same time: that which exists between those colonial relations and national integration (since the existence of the former represents an obstacle to the latter). In other words, national integration may be achieved, not by eliminating the Indian, but only by eliminating him as a colonized being.[72] Mexican indianism has admitted this timidly and not

[72] The term 'national integration' is very ambiguous. The way it is used by Myrdal, for example, referring to its economic aspects, it simply means equality of opportunities (*Cf.* G. Myrdal, 1956: *Solidaridad o desintegración* (Mexico)). When Aguirre Beltrán in: *El proceso de aculturación*, speaks of 'intercultural integration' at the regional level, he rather refers to the homogenization of the cultural differences between Indians and Ladinos, that is, to the predominance of the mestizo culture, which is why we affirm, differing from Aguirre Beltrán, that national integration may be achieved without the disappearance of the 'cultural' Indian.

without some ambiguities. But in this respect it is much more advanced than the rest of the national society. Indianism certainly does not escape the contradictions of national society when, for instance, it is stated that literacy in the Indian language in Chiapas only serves to facilitate the teaching of Spanish, and a series of 'assimilationist' measures (particularly the action of 'acculturation agents' or 'promoters of cultural change') are simultaneously put into practice.

2.5.2 Class Relationships

We cannot over-emphasize that the class character and colonial character of inter-ethnic relations are two intimately related aspects of the same phenomenon. They are separated here only for the purpose of our analysis. Class relationships have developed parallel to and simultaneous with colonial relations and tend to displace them more and more. But the colonial character of inter-ethnic relations impresses particular characteristics upon class relations, tending to stop their development. In this context, class relations mean mutual interactions between persons holding opposed economic positions, independent of ethnic considerations. These relations develop together with the region's economic development. As agricultural production increases, as the market for industrial products expands, as monetary economy develops, and as the labor market expands, colonial relations lose their importance and give way to the predominance of class relations. The latter's development also depends, to a great degree, upon structural factors of national economy and is not the result of decision-making at the regional or local level. At any rate, this development tends to impress upon the class relations between Indians and Ladinos a characteristic mark while the 'feudal' or 'semi-feudal' aspects, so frequently indicated in the literature, tend to disappear.

Consequently, measures for local or community development such as improvement of agricultural techniques, establishment of production co-operatives, etc., may change colonial relations into class relations, but not necessarily so. This transformation can only take place if such developments are accompanied by parallel development of the regional economy as a whole, and particularly of its Ladino metropolis. If such is not the case, the likelihood is that the fruits of local development will enter the traditional socio-economic circuits without modifying the regional structure.

It has already been seen that on certain occasions Ladinos are interested in maintaining colonial relations. There also exist circumstances in which they are interested in strengthening class relationships to the detriment of colonial relationships. This happens particularly with the development of the productive forces: when Ladinos are presented with new opportunities of investment, when they need seasonal labor which can only be obtained through monetary incentives, or when they require non-agricultural labor (for certain manufacturing industries or for construction work in the cities or on the roads); finally, when they need to develop new regional markets and the strengthening of the Indians' demand for manufactured products. The Ladinos' interest in the development of class relations also arises when the agrarian reform manages to really break the land monopoly and when the possession of his own land can turn the Indian back to subsistence farming. In this case, class relations develop particularly through the marketing of crops and the agricultural credit structure.

Under certain circumstances Ladinos may have an interest toward curbing the development of class relations: for instance, when their interests are affected by the establishment of plantations by foreign companies, which modify the status quo by attracting a certain amount of labor and paying higher wages than those which are usual in the region, etc. This has happened in Guatemala. Or, for example, when economic development of the region contributes to the liberation of labor, thus increasing its emigration or at the least its capacity to demand higher salaries, in which case the Ladino latifundists are forced to invest a greater amount of capital in agriculture, and this capital they do not possess.

Indians are also interested in the development of class relationships because these imply the existence of better economic opportunities and of wider alternatives for action. On the other hand, they may be interested in curbing the development of class relations because they tend to destroy the subsistence economy, because they contribute to economic and psychological insecurity and encourage proletarianization and disintegration of Indian culture.

The development of class relations involves new forms of sociability and social organization; there emerge new social categories and new groupings and social institutions. The development of these relations tends to destroy the rigidity of social stratification, to modify its bases (from ethnic characteristics to socio-economic indices) and to encourage ladinization of the Indian.

2.5.3 Social Stratification

Insofar as the regional system of social stratification has only two strata based essentially on ethnic characteristics it tends to maintain the appearance of a colonial situation. At the same time, it tends to change into a clearly defined socio-economic stratification. The already existing stratification among Ladino ethnic groups tends to become extensive to both ethnic groups. Perhaps the day will come when both ethnic groups—independent of their cultural characteristics—will be included into a single stratification system, based exclusively on socio-economic criteria. The old stratification system, based on ethnic characteristics (sometimes called castes) tends to conflict with the development of class relations and the socioeconomic stratification based on them. Thus, for instance, an Indian trader or landowner receives discriminatory treatment from Ladinos who are in a socio-economic situation inferior to his own, while Indian day laborers tend to receive smaller wages than the Ladinos who are in the same position. Among the Ladinos there exists an obvious concern over maintaining the bases of ethnic stratification; especially among the lower strata of the Ladino population, who in this way avoid competing with mobile Indians. This is the same phenomenon as that of the poor whites in the south of the United States and other such cases in other parts of the world.

Social stratification, as we have seen, comprises two aspects: inter-ethnic stratification reflects its colonial past, while Ladino socio-economic stratification, in which Indians are increasingly participating, reflects the development of new class relations, devoid of their ethnic content. The Indians' upward vertical mobility in the socioeconomic scale is accompanied by a certain degree of ladinization, but, as has already been pointed out, not all of the aspects of Indian culture change at the same rate. Development of class relations tends to facilitate the Indian's upward mobility, since an ascent in the socio-economic scale renders the conservation of a low status based upon exclusively ethnic criteria more precarious. Upward mobility, as much in the socio-economic scale as in the shift from the Indian to the Ladino ethnic group, is a function of the transformation of the colonial situation into a class situation.

2.5.4 Ladinization

This process of acculturation of the Indian is hard to place in a structural analysis, since it is used in the literature to refer to processes which are highly varied in content. In a general sense it means the adoption of Ladino cultural elements by individuals or groups (communities) of the Indian ethnic group. Thus, the change in dress, the substitution of folk medicine by scientific medicine, and the change of occupation, to take only three examples, are all part of the process of ladinization. Yet the structural significance of these three examples, taking each by itself, is very different. Without considering for the moment the motivational determinants leading to a change in dress, this by itself has no consequences for the social structure; except if, carried out collectively by the Indians, it should lead to certain changes in the value systems of both ethnic groups, which in turn might influence the systems of mutual action and interaction, thus affecting social structures. But this kind of chain argument does not lead to a better understanding of the phenomena being studied. Of the preceding examples, the second—the shift from traditional medicine to modern medicine—does not by itself represent a Structural change in either. But it may lead to demographic consequences which will have important structural results. Change of occupation, on the contrary, can only be understood within the frame of a structural analysis. The above shows that the concept of ladinization may mean anything from a simple change in the daily use of an object (using a spoon instead of a tortilla to eat soup), up to a complete change of the Indians' life and world view. Within the limits of this essay, concern over the process of ladinization is only meaningful insofar as it has immediate structural implications.

Rodolfo Stavenhagen amidst young Masai warriors in Kenya. *Source* The author's personal photo collection

With displaced indigenous Bedouins in Naqab (Negev) Israel

Chapter 3
Indigenous Peoples: An Introduction (2009)

Abstract This chapter offers a brief introduction to the concept, characteristics and situation of indigenous peoples in the world today. This text is an entry in the Oxford Encyclopedia of Human Rights, published in 2009.

Indigenous peoples are also referred to as natives, aborigines, tribes, autochthons, pueblos originarios (original people), and many other labels that are meaningful at the national or regional levels, such as Indians, First Nations, and Adivasis. Indigenous peoples received little attention in the human rights discourse before the late twentieth century. If acknowledged at all, they were held to belong to some bygone era that had been replaced in modern times by the ubiquitous nation-state, where universal individual human rights are said to prevail.

Thus in human rights terms there world be no need for special attention to be paid to any particular segment of the national society if everybody enjoyed the same universal rights. The exceptions to this widely held paradigmatic view have always been ethnic, national, or religious minorities, for whom special regimes were devised or to whom particular legislation might apply, such as affirmative action, proportional electoral representation, or autonomous regional administration. Indigenous populations have sometimes been included, but most frequently they were ignored in such policies. Early stirrings regarding human rights issues of indigenous peoples took place in a number of countries, but it was not until well alter World War II, when human rights had become an important ingredient of international relations and democratic governance, that indigenous peoples began to organize and lobby for their rights at the domestic and international levels, so that governments and multilateral organizations started to take notice.

This chapter was first published in 2009 as: "Indigenous Peoples", in: David P. Forsythe (Ed.): *Oxford Encyclopedia of Human Rights*, Vol. 3: *Index on Censorship-Minority Rights: Overview* (Oxford–NewYork: Oxford University Press): 17–27. The permission to republish this text was granted on 25 July 2012 by Ms. Mary Bergin-Cartwright, Permissions Manager, Oxford University Press, USA.

R. Stavenhagen, *The Emergence of Indigenous Peoples*, Texts and Protocols 3,
DOI: 10.1007/978-3-642-34144-1_3, © The Author(s) 2013

3.1 Historical Background

The plight of indigenous peoples under colonial domination was documented as early as the sixteenth century by the friar Bartolomé de las Casas, who argued their case at the Spanish court. In the Americas and elsewhere indigenous peoples were frequently victims of massacres, arbitrary resettlement, and various forms of servitude, slavery, or forced labor. During the period of expanding national states (the 'conquest' of the West, the South, or the Great North, according to the circumstances, especially in America), they were usually pushed out of the way by invading forms and cattle ranches and later by agroforestry, plantations, railways, mining, oil rigs, fisheries, and other activities of a growing economy. In the process many indigenous communities around the world lost most of their ancestral lands, traditional territories, sacred forests, coastal areas, and age-long free access to natural resources of all kinds.

The restitution and protection of their homelands are a persistent claim of indigenous communities the world over. The survival and cultural identities of indigenous peoples are closely linked to the concept of Mother Earth, from the Pachamama in the Andean Highlands to the spirit forests and sacred groves of tribal communities in Asia. Global warming and climate change, world pollution, and rapid transformations of land use are affecting their subsistent environments dramatically, as, for instance, the persistent man-made Amazon forest fires or the thinning ice in the Arctic that affects the food chains on which Inuit, Sami, and other northern peoples depend for their living. The highly vulnerable hunter-gatherer communities in the remaining tropical forests of South America, central Africa, and Southeast Asia are particularly affected by these changes, as are the pastoralist herders in the semiarid areas of eastern Africa (Maasai and others). In the more densely populated farming areas of Central and South America and Asia, where most of the world's indigenous peoples were concentrated in the early twenty-first century, the privatization of the land and its resources (forests, waters, minerals) for commercial exploitation as well as major development projects, such as hydroelectric power plants, were putting enormous pressure on indigenous farming communities, leading to increasing rural poverty and emigration to urban centers and other areas of economic growth.

3.2 Who Are Indigenous Peoples?

One of the stumbling blocks to reaching an international consensus on the special character and scope of the human rights of indigenous peoples as well as the specific areas in which their protection may be ensured by state action is the ambiguity surrounding the definition of 'indigenous'. States adopt different definitions in terms of their particular contexts and circumstances. In some countries local terms are commonly used that are not easily translatable; in others the existence of indigenous groups is denied altogether, and therefore their definition becomes even more problematic. In the late twentieth century formal definitions became more common in national legislation concerning rights and issues of indigenous peoples, whereas in other cases such legislation does not provide a definition at all. Another problem concerns the criteria for membership in an indigenous group, nation, or community.

While indigenous peoples the world over share many commonalities, there are also widespread differences among them. In North America, for example, indigenous nations were considered sovereign and distinct by the governments of the United States and Canada well into the nineteenth century, as recognized in numerous treaties. Early twenty-first-century issues concerning their human rights derived to a large extent from violations of these treaties. Legal criteria determine who is Indian in each case. With or without treaties, Indians, just as the Aborigines in Australia, were resettled on reserves or reservations, which in turn as time went on, lost chunks of their land to outside interests.

When most Latin American countries established their democratic constitutional structures in the early nineteenth century, they initiated a process of formal recognition of the indigenous Amerindian population as part of the people of the nation, a process that in practice took more or less time under different conditions. The sparse jungle tribes mainly found in the Amazon basin were subjected to special tutelary regimes and achieved full citizenship rights only much later, whereas some groups remained in voluntary isolation in the early twenty-first century. In contrast, the identity of the Indian peasant farmers in the Andean and Mesoamerican highland areas was considered to be based more on socioeconomic conditions (labor status, access to land property, standards of living) than on ethnic criteria. To be identified as an Indian was taken more as a stage in socioeconomic evolution than as a marker of ethnic identity implying specific rights and obligations. The social science literature on Latin America's Indians during the twentieth century reflects the debates between different theoretical perspectives that frequently conditioned the various social policies or political approaches adopted by a specific administration.

The situation is quite different in Africa, where the notion of 'indigenous people' usually refers to all African populations who were subjected to colonial regimes. Indeed in postcolonial Africa it is often said that 'everybody is indigenous'. The term 'indigenous' as a separate category is frequently rejected altogether with the fear that 'tribalism' may threaten the unity of the state. In 2000 the African Commission on Human and Peoples' Rights established an expert working group on indigenous or ethnic communities in Africa, which took the view that there are indigenous people on the continent whose rights must be protected. The commission recognized the concerns over the use of the term 'indigenous peoples' in the African context, but also considered that within the majority population that is indigenous to the continent some are in a structurally subordinate position to the dominating groups and the state, leading to marginalization and discrimination.

The states of Asia also present different approaches. The various 'tribal' categories used in many countries are equivalent to the concept of indigenous peoples, even when they are not recognized as such officially. This is the case of the Adivasis in India, the Orangasli in Malaysia, the Ainu in Japan, and numerous minority nationalities in China. No such definitional problems arise concerning indigenous peoples in Australia, New Zealand, Scandinavia, or Russia, where a long-standing legal tradition provides clear guidelines regarding the definition of groups and criteria for membership.

The United Nations Commission on Human Rights in 1995 adopted four principles to be taken into account in a definition of indigenous peoples: (1) priority in time with respect to the occupation and use of a specific territory; (2) the voluntary perpetuation of cultural distinctiveness, which may include the aspects of language, social organization, religion and spiritual values, modes of production, laws, and institutions; (3) self-identification as well as recognition by other groups or by state authorities as a distinct collectivity; and (4) an experience of subjugation, marginalization, dispossession, exclusion, or discrimination, whether or not these conditions persist. These principles were incorporated to a greater or lesser degree in a number of national legislations on the rights of indigenous peoples and are used widely in the emerging international law of indigenous peoples at the regional and universal levels. By these criteria there are many thousands of identifiable indigenous peoples in the world, mainly in Asia, Africa, and the Americas, whose populations range from the millions in some cases to several dozen families verging on extinction in other parts. Although there are no hard statistics of the total population of these groups, an informed estimate is around 400 million persons worldwide, divided into many thousands of distinct peoples and communities. Whereas some states regulate individual membership, it has become increasingly accepted that the right to decide who is or is not an indigenous person belongs to the indigenous peoples alone. Membership in indigenous communities implies not only rights and obligations of the individual vis-á-vis his or her group but may also have legal implications involving relations to the state. Some countries, such as the United States, have legislated strict criteria as to membership in indigenous communities. Canada distinguishes between different categories of indigenous or aboriginal people and different kinds of membership that pertain to each of them. In most Latin American countries membership tends to be subjective and flexible based on rather ambiguous cultural criteria, such as mother tongue, family links to specific local communities known to be indigenous, and self-identification. In the late twentieth century constitutional and legislative reforms in Latin American countries attempted to refine the issue of definitions, particularly with regard to land and territorial rights.

3.3 State Policies

A variety of policies regarding the indigenous populations within their territories have been adopted by states in modern times. During the early stages of nation building, indigenous peoples were excluded from the emerging polity altogether and were usually treated with contempt as 'savages' or 'barbarians'. In settler societies they were expelled from their habitats, physically eliminated, or subordinated into a dominant economic and political structure not of their choosing. Special legislation was devised to set indigenous populations apart, generally treating them as not legally competent (denying them citizenship rights, for example) and setting up special government departments to deal with them. In some countries, as in those in Latin America, this task was left to the Catholic Church or other religious missionary organizations. In several British colonies the Crown negotiated or imposed treaties on Indian nations whereby the tribes ceded much of their land and their sovereignty in return for certain limited guarantees and rights. These treaties were later disregarded or abrogated by the governments concerned, as were the Treaty of Waitangi (1840) between the British Crown and a number of Maori tribes in New Zealand and most of the treaties that were signed in Canada and the United States.

3.3.1 Assimilation

Assimilation became the most widespread and common policy of national states with regard to indigenous populations. The Japanese carried out a policy of Japanization of the original Ainu inhabitants as they expanded their hold over the northern island of Hokkaido during the nineteenth century. Despite a number of parliament's (negotiations) between the Spanish colonizers pushing down from the north and the Mapuche peoples of southern Chile that led to a mutually recognized frontier along the Biobio Ríver, the national state pursued an aggressive policy of Chilenization of the Indians during the nineteenth and twentieth centuries. The policies of assimilation denied indigenous peoples their own collective identity by dismantling their social institutions and cultural values in order to incorporate them into an acceptable subordinate position in the dominant society and the expanding capitalist economy. In this process religious conversion-mainly to Christianity-and formal schooling became crucial elements.

At the symbolic level the aim of assimilation was to strengthen the national identity of emerging states as part of nation building. For several generations the Indian children who were forced into Canada's 'residential schools' to be educated by the federal government suffered sustained culture loss, an issue that became a subject of national controversy and remedial measures in later years. Some post-colonial governments of Africa and Asia undertook to 'modernize' and 'integrate' all ethnic minority populations, including those identified as indigenous.

Latin America's official policy of indigenismo (government programs for the development and assimilation of Indian communities), adopted at the *First Inter-American Indianist Congress* in 1940s proposed to integrate the American continents' indigenous populations into the national mainstream.

3.3.2 Multiculturalism

A shift in the major orientation of government policies toward indigenous peoples occurred during the 1980s, when the notion of multiculturalism within a human rights perspective began to challenge the earlier idea of assimilation. While it is always complex to trace the origin of a widely used concept in the humanities and the social policy field, in this case the recognition of the collective rights of indigenous peoples resulted from a change in the general international human rights environment, the depreciation of models of social analysis based exclusively on class and economic interest, the reemergence of ethnic movements and conflicts around the world, and as a result of intense lobbying by indigenous organizations both at the national and international levels, the opening up of the United Nations (UN) system to the long-standing demands of indigenous representatives. Indigenous peoples emerged as new social and political actors in the 1980s, and their rights were incorporated progressively into national and international legal regimes.

3.4 International Human Rights

Formal interest by the UN in the human rights of indigenous peoples began in the early 1980s, when the Economic and Social Council decided to establish a Working Group on Indigenous Populations in the Commission on Human Rights. Its annual sessions were the occasion for several hundred representatives of indigenous organizations from around the world to meet, exchange information, get to know the UN system, and practice indigenous diplomacy in posing their various human rights claims before an international audience. The principal outcome of these debates, alter more than twenty years of negotiations, was the adoption by the Human Rights Council (formerly the Commission on Human Rights) in 2006 and by the General Assembly in September 2007 of the *United Nations Declaration on the Rights of Indigenous Peoples*. A majority of 143 member states voted for it, and only four countries voted against—Australia, Canada, New Zealand, and the United States.

The Permanent Forum on Indigenous Issues, an expert group including indigenous representatives, was formed in 2003 within the UN secretariat to deal with the human rights of indigenous peoples among other areas. The Human Rights

Council created the mandate of special rapporteur on the human rights and fundamental freedoms of indigenous people in 2001, an office held until 2008 by the author of this article. A number of UN human rights treaty bodies also address indigenous rights issues, among others, in their periodic reports and recommendations to states parties.

3.5 The UN Declaration on the Rights of Indigenous Peoples

As a legal international human rights instrument, the UN Declaration on the Rights of Indigenous Peoples complements earlier texts, such as Convention 107 on Indigenous and Tribal Peoples (1957) and its revision, Convention 169 (1989), of the International Labour Organization.

The *United Nations Educational, Scíentific, and Cultural Organization's* (UNESCO) *Convention on Cultural Diversity* (and its earlier declaration on this topic) refers to indigenous peoples, as does the Biological Diversity Convention, among others. In the Western Hemisphere the *Organization of American States* was still preparing an *American Declaration on the Rights of Indigenous Peoples* in 2008. A proposal to do likewise was tabled in the *Organization of African Unity*.

The UN Declaration on the Rights of Indigenous Peoples did not create any new rights; it built on earlier well-established universal human rights instruments and related these to the specific circumstances and needs of indigenous peoples. In its articles it refers to the individual rights of indigenous persons and also to the group rights of indigenous communities and peoples. This approach signaled a break with the traditional international perspectives on the rights of minorities and indigenous peoples that concentrate exclusively on individual rights by referring to 'members' of rather than the group as a whole. Article 3 of the declaration recognizes that indigenous peoples have the right to self-determination, in line with Article 1 of the two international human rights covenants of 1966, in other words, that the rights of peoples also apply to indigenous peoples.

During the negotiations leading up to the UN General Assembly vote, this right was sharply contested by a number of states that perceived it as a threat to their territorial integrity. The question of who is actually the bearer of this right and under what circumstances it can be exercised was still open and controversial) in the early twenty-first century. Usually, it is interpreted as the right to internal autonomy or self-government (Article 4) and is related to the right of indigenous peoples to maintain and strengthen their distinct political, legal, economic, social, and cultural institutions (Article 5 and Article 20). Other articles refer to the rights of indigenous peoples to their own cultures, languages, education, and media; not to be subjected to forced assimilation or destruction of their cultures; and to free, prior, and informed consent in relation to legislative and administrative measures that may affect them. The right to determine and be actively involved in development is acknowledged in Article 23.

The controversial Article 26 states that indigenous peoples have the right to the lands, territories, and resources that they have traditionally owned, occupied, or otherwise used or acquired.

3.6 Indigenous Peoples Right to Land, Territory, and Natural Resources

The land rights issue is one of the principal sources of conflict among indigenous peoples, private landholders, and government authorities. From time immemorial indigenous peoples maintained a special relationship with the land, their source of livelihood and sustenance and the basis of their very existence as identifiable territorial communities. The right to own, occupy, and use land is inherent in the self-conception of indigenous peoples, and generally it is vested in the local community, the tribe, the indigenous nation or group. For economic productive purposes this land may be divided into plots and used individually or on a family basis, yet much of it is regularly restricted for community use only (forests, pastures, fisheries, and so forth), and the social and moral ownership belongs to the community.

Such tenure arrangements are sometimes recognized in the national legal system, but just as often governments promote the transfer of communally held indigenous lands to other landholders, a process that began during the colonial period in many countries and intensified during postcolonial times. The breakup of indigenous agrarian communities in the nineteenth century was one of the factors leading to the Mexican Revolution of 1910.

Mapuche communities in southern Chile assertively oppose the concentration of their ancestral lands in private hands, a process actively promoted since the military dictatorship of the 1970s. In former British colonies where the Crown took possession of native lands and territories, ancestral title to land was often extinguished through legislation, administrative process, or judicial sentence to be transferred to private ownership. In Canada and New Zealand, among others, aboriginal communities persistently engage in defending their ancestral rights to these lands by 'legislation, litigation, or confrontation', as a local saying among First Nations goes. The Masai pastoralists of Kenya resist the government's efforts to persuade them to become individual landowners on their traditional communally held grazing areas, arguing that this would destroy their subsistence economy and hence their culture and lifestyle. In numerous Asian countries as well, indigenous peoples are being dispossessed of their land by large private or corporate economic interests, usually with government support. Despite some (generally weak) protective legislation, the loss of indigenous lands and the extinguishment of aboriginal title proceed relentlessly, raising concerns about the human rights consequences for indigenous peoples. In a number of countries these problema have led to rural unrest, social conflict, and sometimes violence.

Indigenous communities and human rights organizations frequently work together to protect the Lands to which they have a claim according to international and national legal standards. In a landmark case in 2001 the Inter-American Court of Human Rights decided in favor of the Awas Tingni indigenous community against the government of Nicaragua, which had failed to protect their customary land rights. Although similar judgments have been made by the courts elsewhere, they are exceptional cases, because indigenous communities do not have easy access to the judicial system and in a number of countries these remedies are not available to the indigenous people at all. Even when favorable laws are on the books, they are not always implemented. In some countries legislation undermines traditional communal or tribal holdings and opens the way to their dispossession by third parties or other private or corporate interests. Violations of indigenous land rights within the framework of national development programs are a major source of social tensions in a number of countries.

Indigenous communities maintain historical and spiritual links with their homelands, geographical territories in which society and culture thrive and that therefore constitute the social space in which a culture can reproduce itself from generation to generation. Too often, this necessary spiritual link between indigenous communities and their homelands is misunderstood by no indigenous persons and is frequently ignored in existing land-related legislation. Although the recognition of indigenous territorial rights appears to be necessary for the full protection of the human rights of indigenous peoples (an argument wielded by indigenous organizations), governments often fear that such recognition might undermine the unity and integrity of existing states. Nevertheless, in a number of states such rights have indeed been legislated, and experience suggests that national unity is not threatened by these developments.

Cases in point are Nunavut in Canada and the five indigenous *comarcas* (administrative territories) of Panama. Brazil, Colombia, and Ecuador, among other Latin American countries, have legislated on specific territories for indigenous peoples. When indigenous territories are rich in resources coveted by outside interests, local communities tend to suffer considerable pressures that frequently lead to serious social conflicts.

3.7 Indigenous Peoples the Struggle Over Resources

The struggle over resources for social and economic development is directly related to the right of indigenous peoples to land and territory, as recognized in Convention 169 of the International Labour Organization (Article 15) and in Articles 29 and 32 of the UN Declaration on the Rights of Indigenous Peoples. Frequently, such resources are extracted or developed by other interests (oil, mining, logging, fisheries, and so forth) with little or no benefits for the indigenous communities that occupy the land. In countries such as Russia, Indonesia, or Cambodia, indigenous communities engage with governments and the private sector to find common solutions to these emerging and growing problems. The UN proposed a global compact involving governments and private enter-prise to further human rights-based sustainable development.

The compact is based on two sound principles: businesses should support and respect the protection of internationally proclaimed human rights, and they should make sure that they are not complicit in human rights abuses. This is what indigenous peoples have been struggling for.

3.8 Impacts of Development

Indigenous peoples live mainly in rural environments. They are able to maintain their community lifestyles and traditional cultures only where their habitats have been spared major upheavals resulting from rapid economic and ecological transformations. This is not usually the case when national governments, large corporations, and multilateral financing agencies turn their attention to so-called undeveloped regions to extract natural resources; establish plantations and industrial plants; develop tourist activities, ports, communication hubs, or urban centers; or build transportation networks, multipurpose dams, military bases, or toxic waste dumps. Wherever such developments occur in areas occupied by indigenous peoples, it is likely that those communities will undergo profound social and economic changes that are frequently not well understood, much less foreseen, by the authorities in charge of promoting them. Large-scale development projects inevitably affect the conditions of living of indigenous peoples. Sometimes the impact is beneficial, very often it is devastating, but it is never negligible.

When government agencies and development corporations plan and execute major construction projects, the concerns of indigenous peoples, who are seldom consulted on the matter, tale a back seat to overriding 'national interests' or to market-driven business objectives aiming at developing new economic activities, maximizing productivity and profits. Due to the environmental degradation and pollution they produce, major development projects often entail serious physical and mental health hazards for indigenous peoples, who argue that major development projects that do not tale into account their fundamental interests entail violations of their basic human rights. The World Commission on Dams found that indigenous and tribal peoples suffered disproportionately from the negative impacts of large dams while often being excluded from sharing in the benefits.

Often these projects entail involuntary displacements and resettlement of indigenous communities that happen to lie in their way. As a result, violations of civil, political, economic, social, and cultural rights occur with increasing frequency, prompting indigenous peoples to launch major protests or resistance campaigns to bring public attention to their plight, besides engaging the judicial process or appealing for administrative redress as well as lobbying the political system. Reports from many countries, among them Australia, Bangladesh, Botswana, Chile, Kenya, the Philippines, Malaysia, India, Thailand, Ecuador, Peru,

Japan, the United States, and Russia, point to the persistence of major human rights violations involving indigenous communities affected by such development projects.

Evictions or involuntary displacements are a common feature and have been disapproved by the United Nations Committee on Economic, Social, and Cultural Rights. In the late twentieth century this situation began to change as multilateral agencies, national governments, and the private sector took a new interest in indigenous concerns. The International Labour Organization's Indigenous and Tribal Peoples Convention 169 and the UN Declaration on the Rights of Indigenous Peoples recognized the right of indigenous peoples to decide their own priorities for the process of development. At numerous international conferences these rights were reaffirmed, notably at the Rio Earth Summit (1992) and the Johannesburg World Summit on Sustainable Development (2002).

The World Bank, which finances many of the world's major development schemes, adopted Operational Policy 4.10, aiming to protect the interests of indigenous peoples in these projects. None have been more involved in these important issues than indigenous peoples themselves, who have raised their human rights concerns with respect to development projects time and again at international conferences and who continue to lobby major international agencies and private corporations.

3.9 Indigenous Peoples Legislation and Justice

The rights of indigenous peoples, as communities and persons, can be excluded, ignored, or protected by national legal systems. When indigenous peoples are specifically excluded from human rights legislation (for example, when they have been formally declared to be wards of the state and therefore not free actors in a national polity), then it can be said that a serious human rights protection gap exists, even when other kinds of protection mechanisms are available. When they are simply ignored as a specific category of rights holders, then one would expect the rights of indigenous persons to be sufficiently covered by other universal human rights instruments as legislated domestically. If this does not occur, then one would expect the indigenous peoples-like everybody else-to have access to general existing human rights protection mechanisms. If, however, there is specific legislation addressing the rights of indigenous peoples-as communities and as individuals-in a given state, then one should expect adequate implementation mechanisms to be put in place to ensure that these rights are fully enjoyed by such communities and persons. During the last two decades of the twentieth century a number of countries adopted such legislation to a greater or lesser degree, an incomplete process that opened up new spaces for the protection of the rights of indigenous peoples. The question in the early twenty-first century is to what extent this legislation is actually being implemented and how it impacts the human rights of indigenous peoples. To be sure, the situation varies considerably from one country to another, yet the evident points to the existence of a serious implementation gap between the normative framework and administrative, legal, and political practice. This divide between form and substance constitutes an ongoing violation of the human rights of indigenous peoples.

3.9.1 Administration of Justice

Administration of justice is one of the more problematic arcos where this implementation gap is particularly serious. Indigenous peoples have been the historical victims of persistent patterns of denial of justice, and they struggled hard to overcome the structural injustices of which they are victims. The First Nations of Canada have obtained some notable successes in litigating for their ancestral treaty rights; so have the Sami in Norway, the Aborigines in Australia, the San in Botswana and in South Africa, the Maya in Belize, and the U'wa in Colombia, among many others. But litigation is a drawn—out and expensive process not always available to indigenous peoples in other countries that have different legal traditions. Confrontation and social mobilization occur in places where ordinary legal and political processes do not work for indigenous peoples. Some national laws maintain the alienation and exclusion of indigenous peoples from the justice system altogether. For example, Nepal was officially a Hindu kingdom and Nepali the official language, granting no recognition to its indigenous people who suffered discriminatory legislation. In 2008, after a long and bloody civil war, the hereditary monarchy of Nepal was abolished and the country was declared a republic, opening a space for a new deal for Nepal's indigenous peoples. Early twenty-first century reports indicated that in the Russian Federation the rights of indigenous peoples were not protected despite guarantees to the contrary in the 1999 Federal Law on the Guarantees of the Rights of Indigenous Numerically Small Peoples of the Russian Federation. The main problem appeared to be the lack of implementation of the federal law at the regional and local levels, a concern expressed repeatedly by international bodies and experts.

A complex situation that indigenous peoples encounter frequently is the incompatibility between human rights legislation pertaining to indigenous peoples and other sectoral laws (such as legislation regarding the environment or the exploitation of natural resources or the titling of private landholdings). When asked to rule on competitive claims over such issues, courts do not always protect the rights of indigenous communities; and legislative bodies have been slow to correct existing imbalances when the human rights of indigenous peoples are involved. The widespread lack of access to the formal justice system due to ingrained direct or indirect discrimination against indigenous peoples is a major feature of the human rights protection gap. Often, this only reflects the physical isolation and lack of means of communication in indigenous arcas, but it also occurs when public resources are simply insufficient to build up an effective judiciary designed for the needs of indigenous communities. More seriously, it

Rodolfo Stavenhagen with Quechua women in Ecuador. *Source* Personal photo collection of the author

may signal the facts that the official legal culture in a country is not adapted to deal with cultural pluralism and that the dominant values in a national society tend to ignore, neglect, and reject indigenous cultures. Numerous studies document bias or discrimination suffered by indigenous persons in the justice system, particularly in the criminal justice area, where women, youth, and children are particularly disadvantaged. Of special concern is the overrepresentation of indigenous peoples in criminal proceedings and prisons, as indicated by official reports on Australia, Canada, Mexico, the United States, and New Zealand, among others. Studies also show overcrowding and substandard conditions in prisons, where indigenous peoples are detained. Anecdotal evidence points to a similar situation in other countries where official statistics are unavailable.

The language used in judicial proceedings is often a problem for the adequate protection of the rights of indigenous peoples. In countries where indigenous languages are not officially recognized, this may be a major disadvantage. Judicial personnel (including judges) may not have sufficient cultural training or sensitivity to deal with the human rights issues of indigenous peoples. Some countries promote the use of 'court facilitators' to help indigenous peoples in judicial proceedings, and special attention may be paid to indigenous children, who are particularly vulnerable to practices of discrimination. Traditional methods of restorative justice are a means of responding positively to offenses committed by children, in accordance with human rights principles. Indigenous traditions can be a means of dealing with children in conflict with the law in a constructive manner, without resorting to punitive justice, hence favoring their reintegration into the community.

The obstacles indigenous peoples face in the justice system are merely symptoms of a larger picture of complex social problems related to a history of discrimination, racism, marginalization, and social exclusion, including poverty and unemployment, which is often expressed through alcoholism and drug abuse, homelessness, and violence. Discrimination against indigenous peoples in the justice system is a widespread occurrence. While it is often related to the personal prejudices and subjective attitudes of judges, magistrates, attorneys, prosecutors, and government officials, it is even more related to systemic rejection of indigenous cultures and identities. The justice system does no more than express the dominant values of a society, and when these are biased against indigenous peoples (as is so often the case), the courts tend to reflect them. Only in the late twentieth century, and to a great extent as a result of developments in the international arena, did the atmosphere begin to change.

The justice system has been used on occasion to criminalize social protest and the legitimate demands of indigenous organizations in defense of their rights. This trend appeared in two guises: the application of emergency legislation, such as antiterrorist laws, and accusing activists of common misdemeanors (such as trespassing) to repress social protests. Many examples have been documented in numerous countries by human rights observatories. Human rights abuses occurred not only during states of emergency or in authoritarian nondemocratic regimes but also within the framework of the rule of law in open, transparent societies. Rights abuses committed against indigenous peoples often happen in the context of collective action by indigenous organizations to press their social claims. Private vested interests and beleaguered authorities belonging to local power structures often use the law to dismantle such movements by penalizing prominent leaders either through the application of common criminal statutes and regulations or by invoking politically motivated antiterrorist legislation.

3.9.2 Customary Law

The UN Declaration on the Rights of Indigenous Peoples provides for the right of indigenous peoples to their own customary justice systems. The no recognition of indigenous law is part of a pattern of the denial of indigenous cultures, societies, and identities and the difficulty that modern nation-states have in recognizing their own multicultural composition. In many countries a monist conception of national law prevents the adequate recognition of plural legal traditions and leads to the subordination of surviving customary legal systems to one official norm. Given existing patterns of discrimination, many indigenous peoples distrust the national justice system and ask for greater control over family, civil, and criminal matters. This reflects questions relating to self-government and self-determination. To remedy the many injustices and indignities that indigenous peoples suffer in the justice system, alternative ways of dispensing justice and solving social conflicts have been attempted in numerous countries. An example is Greenland Home Rule, where the justice system, although still administered by the Danish authorities, is responsive to the standards and values of Greenlandic society and traditional Inuit legal practices and customary law, with a strong emphasis on resocialization and the principle of extensive lay participation in the administration of justice. The Navaho of the United States set up their own Navaho courts that obtained some remarkable results locally. Countries that have been able to incorporate respect for customary indigenous law in their formal legal systems find that justice is handled more effectively, particularly when dealing with civil and family law but also in certain arcas of criminal law, so that a sort of legal pluralism appears to be a constructive way of dealing with diverse legal systems based on different cultural values.

3.10 Indigenous Peoples Right to Education and Culture

Although the right to education is universally recognized, indigenous peoples have not exercised it fully. The degree of illiteracy, poor academic achievement, and poor school attendance, especially at the middle school and higher levels, tends to be higher among indigenous peoples than among the rest of the population. While some countries are making a major effort to improve education levels in indigenous communities, many obstacles impede their access to education. Economic, social, and cultural factors can make it hard for children to take part in educational activities in rural areas, where schools generally lack adequate facilities, budgets, and human, technical, and institutional resources. In urbanized arcas, where indigenous families often coexist with members of majority ethnic groups, the specific needs of indigenous children usually go unattended. Formal education provided by the state or religious or private groups can be a double-edged sword for indigenous peoples. Whereas it often enables indigenous children and youth to acquire knowledge and skills that allow them to move ahead in life and connect with the broader world, it can also be a means of forcibly changing and in some cases destroying indigenous cultures, especially when teaching programs, curricula, and teaching methods come from other societies that are removed from indigenous cultures. A case in point is boarding schools for indigenous children, which in many cases played an important role in ensuring access to and continued attendance at school while also providing, where necessary, food and health—care services, especially when the population was scattered and communication was difficult. However, these institutions were also usually successful in separating whole generations of indigenous students from their cultural roots and very often from their families, doing irreparable harm to the survival of indigenous cultures and societies, a process described by First Nations in Canada as one of ‘generational loss’.

The main obstacles to full enjoyment of the right to education have been assimilationist models of education and the education system's ignorant of or failure to appreciate indigenous languages and cultures. Although large segments of the indigenous populations have indeed become assimilated in many countries, the underside of this process has been the transformation and ultimate disappearance of many indigenous cultures. Over time a great many indigenous languages have continued to vanish around the world. Furthermore, the educational curricula of many countries often reflected indigenous cultures in an inappropriate and disrespectful way, which further contributed to discrimination and prejudice against indigenous peoples in society in general. The academic outcomes of and public spending on indigenous education in the early twenty-first century were on the average considerably lower than for the rest of the population. Comparative data show that school attendance of indigenous girls is generally lower than that of indigenous boys, often as a result of cultural discrimination within their own communities, a situation complicated in certain regions (such as East Africa) by the traditional practice of female genital mutilation. There is also evident of various types of discrimination against indigenous children in schools, particularly when they live alongside no indigenous populations and especially in urban centers, where recent indigenous immigrants live in precarious conditions.

Slowly the situation as described above has been changing, and in the early twenty-first century several countries officially recognized indigenous cultures and agreed on the need for bilingual and intercultural education. Indigenous peoples demand the recognition of their right to education and call for school curricula that reflect cultural differences, include indigenous languages, and contemplate the use of alternative teaching methods. Intercultural bilingual education faces many hurdles, from the lack of adequately trained bilingual teachers to problems in developing the appropriate teaching materials and methods and the need to involve indigenous communities in the design and operation of their own educational institutions at all levels. Progress is being made in this area in many countries, from preschool to higher education, albeit with mixed results. There is a tendency to focus on continuing education as an alternative in the area of indigenous education. UNESCO and educational specialists in many countries recognize that formal and non formal indigenous education, adapted to local cultures and values, is the best way to ensure the right to education; it does not mean shutting out the outside world or ignoring the challenges posed by national societies or the global economy but is in fact viewed by indigenous communities themselves as a necessary tool for the full personal, social, and cultural development of aboriginal peoples. Numerous indigenous communities from Alaska to Peru, from India to New Zealand, have taken multiple local initiatives to preserve their languages and cultures through alternative educational programs. In all cases they help save the traditional knowledge of the aboriginal culture, enhance cultural pride and identity among young people, strengthen ties to the land and the environment, and offer indigenous youth an alternative view of the future.

Indigenous cultures have been trivialized and reduced to picturesque photo opportunities through the global commercial consumer economy. Indigenous peoples everywhere demand respect for their values, worldviews, and cultures and look for ways and means to preserve, protect, and develop these cultures on their own terms. The UN Declaration on the Rights of Indigenous Peoples recognized this as a human right of indigenous peoples. The implementation of cultural rights requires not only access to justice but also adequate public and private cultural policies. Among intergovernmental organizations, UNESCO has promoted cultural diversity as part of an emerging global ethic by means of educational and cultural policies (including communication and information). The protection and preservation of the cultural heritage of all peoples, both tangible (monuments, artworks) and intangible (oral literature, traditions, music), must be an essential ingredient of this global ethic. Indigenous peoples, who have suffered the destruction of their cultures (ethnocide) over many centuries, require adequate public policies to redress the imbalance, thus enabling them to achieve full human development in accordance with their own cultural values and worldviews. An important issue for indigenous peoples is the recognition of the right to intellectual and cultural property. In large parts of the world indigenous peoples are the owners of a vast amount of traditional knowledge, built up from tune immemorial, concerning nature, plants, animals, skills, and techniques of many sorts that are rapidly being swept away by the tsunami of modern society. As important as their possible disappearance is the fact that much of this knowledge is being appropriated, patented, and commercialized by private businesses without die consent of the peoples concerned. Indigenous peoples lobbied effectively at the international level in the late twentieth century for the protection of their intellectual property rights. The World Trade Organization and the World Intellectual Property Organization took up these issues in their ongoing work, but effective intentional legislation (and of course domestic legislation as well) tor the protection of these rights were only at their initial stages in the early twenty-first century.

3.11 Levels of Living and Socioeconomic Development

Research in many countries found that the socioeconomic and human development indicators among indigenous peoples were consistently lower than those among the no indigenous population and that indigenous women and girls showed lower indicators than the male indigenous population. This held across the board for health, education, housing, nutrition, employment, income, access to social services, and other items. National statistical information on these topics is not always disaggregated by race or ethnicity, which makes it difficult to assess the hill scale of the persistent inequality between indigenous and no indigenous populations in many countries. A number of states do not produce such information by design and concentrate their social policies generally on the objective of combating poverty or 'closing the gap' between higher and lower income populations. Indigenous peoples, however, demand that information on levels of living and social services be disaggregated so that public policies may become focused on the specific needs of their communities. Some government services (as well as non-governmental institutions and international agencies) are developing methodologies to concentrate on the crucial differences between the situations of indigenous peoples and other sectors of the population so as to help construct adequate public responses.

These considerations apply equally to urban-based indigenous peoples who do not live in relatively isolated, differentiated territorial communities. Ethnicity-based inequality in the urban environment also affects indigenous peoples unfavorably, but there the public response requires other means, such as the careful monitoring of nondiscrimination legislation or affirmative action in employment, schools, housing, and so forth. Indigenous identities in the urban environment relate not so much to rural-based traditional communities as to the emergence of social and political movements in which ethnicity is not only a marker of identity but also a political statement often used in electoral processes and political negotiations. A case in point was the election of Bolivia s first indigenous Aymara president in 2006.

Bibliography

African Commission on Human and Peoples' Rights, 2005: *Report of the African Commission's Working Group of Experts on Indigenous Populations/Communities* (Banjul, Gambia: African Commission on Human and Peoples' Rights).

Anaya, S. James, 2004: *Indigenous Peoples in International Law* (Oxford - New York: Oxford University Press).

Belgrave, Michael, Merata Kawharu, and David Williams (Eds.), 2005: *Waitangi Revisited: Perspectives on the Treaty of Waitangi* (New York: Oxford University Press).

Brölmann, Catherine, Rene Lefebre, and Marjoleine Zieck (Eds.), 1993: *Peoples and Minorities in International Law* (Boston: Martinus Nijhoff).

Chakma, Suhas, and Marianne Jensen (Eds.), 2001: *Racism against Indigenous Peoples* (Copenhagen, Denmark: International Work Group for Indigenous Affairs).

Cook, Curtis, and Juan D. Lindau (Eds.), 2000: *Aboriginal Rights and Self-Government: The Canadian and Mexican Experience in North American Perspective* (Montreal - Ithaca, N.Y.: McGill-Queen's University Press).

Crawford, James (Ed.), 1988: *The Rights of Peoples* (Oxford: Clarendon).

Dahl, Jens, Jack Hicks, and Peter Hull (Eds.), 2000: *Nuncivut: Inuit Regain Control of Their Lands and Their Lives* (Copenhagen, Denmark: International Work Group for Indigenous Affairs).

Falk, Richard A., 2000: *Human Rights Horizons: The Pursuit of Justice in a Globalizing World* (New York: Routledge).

Gilbert, Jeremie, 2006: *Indigenous Peoples' Land Rights under biter- national Law: From Victims to Actors* (Ardsley, N.Y.: Transnational).

Hanke, Lewis, 2002: *The Spanish Struggle for lustice in the Conquest of America* (Dallas, Texas: Southern Methodist University Press).

International Labour Organization, 1999: *Indigenous Peoples of South Africa: Current Trends* (Geneva, Switzerland: International Labour Organization).

Kymlicka, Will, 1995: *Multicultural Citizenship: A Liberal Theory of Minority Rights* (Oxford: Clarendon).

Läm, Maivän Clech, 2000: *At the Edge of the State: Indigenous Peoples and Self-Determination* (Ardsley, N.Y.: Transnational).
Nicholas, Colin, 2000: *The Orang Asli and the Contest for Resource: Indigenous Politics, Development, and Identity in Peninsular Malaysia* (Copenhagen, Denmark: International Work Group for Indigenous Affairs).
Permanent Council of the Organization of American States, Committee on Juridical and Political Affairs. Working Group to Prepare the Draft American Declaration on the Rights of Indigenous Peoples; at: <http://www.oas.org/council/ CAJP/Indigenous.asp > .
Perry, Richard. J., 1996: *From Time Immemorial: Indigenous Peoples and State Systems* (Austin: University of Texas Press).
Psacharöpoulos, George, and Harry Anthony Patrinos (Eds.), 1994: *IndigepamPeople and Poverty in Latin America: An Empirical Analysis* (Washington, D.C.: World Bank).
Rodriguez-Pmero, Luis, 2005: *Indigenous Peoples, Postcolonialism, and International Law: The IlO Regime (1919-1989)* (Oxford - New York: Oxford University Press).
Roy, Rajkumari Chandra Kalindi, 2000: *Land Rights of the Indigenous Peoples of the Chittagong Hill Tracts, Bangladesh* (Copenhagen, Denmark: International Work Group for Indigenous Affairs).
Sieder, Rachel (Ed.), 2002: *Multiculturalism in Latin America: Indigenous Rights, Diversity, and Democracy* (New York: Palgrave Macmillan).
Sigler, Jay A., 1983: *Minority Rights: A Comparative Analysis* (Westport, Conn.: Greenwood).
Stamatopoulou, Elsa, 2007: *Cultural Rights in International Law: Article 27 of the Universal Declaration of Human Rights and Beyond* (Boston: Martinus Nijhoff).
Stavenhagen, Rodolfo, 2002-2007: *Annual Reports to the Human Rights Council on the Situation of Human Rights and Fundamental Freedoms of Indigenous People* (New York: United Nations document series).
Thornberry, Patrick, 2002: *Indigenous Peoples and Human Rights* (Manchester, U.K.: Manchester University Press).
Tomuschat, Christian (Ed.), 1993: *Modern Law of Self-Determination* (Boston: Martinus Nijhoff).
United Nations General Assembly, 2007: *UN Declaration on the Rights of Indigenous Peoples.* UN Doc. A/61/L.67.
Van Cott, Donna Lee (Ed.), 1994: *Indigenous Peoples and Democracy in Latin America* (New York: St. Martin's Press).
Warren, Kay B., and Jean E. Jackson (Eds.), 2002: *Indigenous Movements, Self Representation, and the State in Latin America* (Austin: University of Texas Press).
Williams, Robert A. Jr., 1990: *The American Indian in Western Legal Thought: The Discourses of Conquest* (New York; Oxford University Press).
Williams, Robert A., Jr., 1997: *Linking Arms Together: American Indian Treaty Visions of Law and Peace, 1600-1800* (New York: Oxford Uhiversity Press).
Xanthaki, Alexandra, 2007: *Indigenous Rights and United Nations Standards: Self-Determination, Culture, and Land* (Cambridge, U.K.: Cambridge University Press).
Young, Crawford (Ed.), 1993: *The Rising Tide of Cultural Pluralism: The Nation-State at Bay?* (Madison: University of Wisconsin Press).

Chapter 4
The Return of the Native: The Indigenous Challenge in Latin America (2002)

Abstract Some years after Cortés conquered Tenochtitlan almost five centuries ago, the chronicler Bernal Diaz del Castillo wrote that had it not been for Malintzin, an Indian woman who served Cortés as interpreter, the conquest of the fabulous Aztec empire might not have taken place. Malinche, as she came to be called, has since been portrayed in Mexico's nationalist historical accounts as a traitor to her people and the term *malinchismo* became synonymous with kowtowing to foreign interests and selling out one's country.

But times have changed: recent appraisals have re-written the story and now Doña Marina, as she was baptized before Cortés gave her in marriage to one of his followers, is hailed as the first exponent of an inter-cultural dialogue that has lasted for half a millennium, a multicultural heroine before her time, who was able to navigate her way among different languages, civilizations, and religions better than any of the accompanying males. She has been turned by some admirers into a feminist icon, the advisor of warrior-chiefs, the interpreter of conflicting interests, the mediator between clashing ideologies, and–why not?–the primal mother of a new race: the *mestizos*, who came to dominate the history of Latin America in later centuries. Times have changed indeed.

Indian heroes and Indian villains have populated Latin America's symbolic universe for five hundred years. Whereas Tupac Amaru rose against the Spanish

---

This text is based on The John Brooks Memorial Lecture I presented in The Chancellor's Hall, Senate House, University of London, on 15 March 2002. This paper was published by the Institute of Latin American Studies of the University of London as Occasional Paper No. 27. (This chapter was first published in 2002 as: *The Return of the Native: the Indigenous Challenge in Latin America* Occasional Papers No. 27 (London: University of London, Institute of Latin American Studies). The permission to republish this text was granted on 20 July 2012 by Professor Linda Newson, Director, Institute for the Study of the Americas. School of Advanced Study, University of London, London, UK.).

DOI: 10.1007/978-3-642-34144-1_4,

overlords in Peru in the late eighteenth century and attempted to reconstruct the Inca kingdom (an initiative for which he was drawn and quartered), Benito Juarez, a son of the Zapotec nation as he described himself, became president of Mexico in the mid-eighteen hundreds, fought the French empire, and has since been revered as one of nationalist Mexico's great civic heroes. He also decreed the disappearance of the communal lands and territories of Indian communities in the name of progress, sentencing most of the country's peasant population to ruin, poverty and despair.

In 1992 the world commemorated the fifth centenary of Columbus' fateful voyage in 1492, and the event gave rise to acrimonious international debates. The Hispanophiles wished to celebrate the anniversary of the 'Discovery of America' and Spain's 'civilizing mission', whereas the Hispanophobes denounced the European invasion which resulted in the genocide of millions of natives in the Americas. Historians waged academic battles over the matter, and the media jumped gleefully into the fray. A compromise was reached to name the event the 'Encounter of Two Worlds', which gave satisfaction to no one but allowed the United Nations to solemnly celebrate the occasion. Indigenous peoples, who had hardly been consulted about the matter, spoke prophetically of '500 Years of Resistance' and succeeded in having the United Nations proclaim the International Year (1993) and two International Decades of Indigenous Peoples (1995–2014).

Nineteen-ninety-two is also the year Rigoberta Menchú was awarded the Nobel Peace Prize for her contribution to the struggle for peace and human rights carried out under conditions of great adversity in Guatemala. Her celebrated memoir, I, Rigoberta Menchú, became a world best seller and placed the desperate situation of the Maya people in the limelight of international attention. Yet a few years later a storm broke over Rigoberta and her testimonio, when an American anthropologist challenged the factual accuracy of her account. The controversy fueled a minor culture war in American academe and was eagerly picked up by those who were determined to delegitimize the claims of Guatemala's Indians to justice and dignity. What right had a semi-literate young Maya woman to question official truths? In a curious twist of logic, the victim became a culprit and her supporters dupes. In 2007 Rigoberta Menchú, now an experienced and well known actor on the international scene, decided to become a candidate for the presidency of Guatemala.

From Malinche to Menchú, the natives of Latin America have always been seen as a 'problem' by members of the dominant society, who typically refuse to recognize that it is this very society that constitutes an unresolved 'problem' for millions of indigenous people in the region. Yet in the tangled web of fact and fiction, legends and lies, memory and myth, rebellion and resistance, indigenous peoples have re-emerged in recent decades as new historical subjects, assertive actors in those fragile and incomplete Latin American democracies that brave the tempests of globalization.

Whilst during the colonial period indigenous peoples had a fixed status in society, ever since the emergence of the region's independent republics, they have been involved in an uneasy and ambiguous relationship with the State and its institutions. As Latin American countries enter the new millennium, they are now faced with the task of reassessing this relationship in the light of the new emerging Indian identities that articulate old grievances and express new demands. The once fashionable theories of social change, modernization and nation-building which dominated social science thinking for over a half-century are now being challenged by the new social movements of indigenous peoples and their developing political ideologies. Will the new millennium bring redemption at last or are we only witnessing one more of those passing social trends that rise and wane in response to external circumstances? Let us attempt to fill in the picture on this amazing development.

Ever since the Europeans first came to the shores of the American continent to conquer and settle, indigenous peoples suffered discrimination, exploitation and racism. Much of the colonial wealth of Europe was based on the use of Indian labor in the mines and fields. Indigenous cultures were destroyed or subordinated to the dominant Hispanic Catholic mold. In some parts of the continent, widespread physical destruction of indigenous societies took place, which is today referred to as genocide and constitutes a crime against humanity. To be sure, the Spanish colonial empire adopted certain measures for the protection of its native vassals, yet they were soon decimated as a result of military conquest, ecological destruction, forced labor and the lethal diseases such as smallpox and measles introduced by the invaders, against which they had no immunity.

The nineteenth century brought independence and a new legal and political system in which Indian populations in most countries were recognized as equal citizens. Nevertheless, the expansion of agrarian capitalism and the modernization of the economy did not bring them many benefits. On the contrary, numerous indigenous communities lost their lands and were forced to provide servile labor on the large estates. Despite holding legal citizenship rights, they were in fact excluded from equal participation in the economic, social and political system. Special legislation often placed them at a disadvantage in relation to the rest of society and in some countries, Indians were treated as minors and legally incompetent until recently.

When the Spanish-American republics achieved their political independence in the early nineteenth century most of them were populated by a majority of Indians, but the power holders were the *criollo* elites, the direct descendants of the Spanish colonial ruling class. Indians remained, as it were, at the bottom of the heap. Indian oppression in the new republics was two-fold. On the one hand, the landowning oligarchies that spread out and consolidated their economic power during the nineteenth century reaped the benefits of the privatization of Crown lands, ecclesiastical estates and traditional holdings of the Indian communities. Soon the remaining Indians were pushed into frontier areas, inaccessible mountain ranges, arid wastelands and impenetrable jungles, while the new *latifundistas* and *hacendados* (large estate owners), and in later years waves of immigrant settlers, took over the best acreages and pastures. In some areas the land was cleared *manu militari* in genocidal ‘pacification’ campaigns. Generations of Indian peasants were forced into peonage and forced labor, and eventually into rural migration circuits and out-migration, a process that has not yet ended.

By the twentieth century, micro subsistence holdings and landlessness had become characteristic of the Indian peasantry, leading to agrarian uprisings and revolutions and to multiple experiments with land reforms. The first such reform came into being after the Mexican revolution (1910–1917) and it transformed the rural landscape in that country. Massive land transfers from the estate sector to peasant communities took place especially during the decades of the nineteen thirties. A later experiment in land reform was undertaken after the 1952 political revolution in Bolivia. A progressive democratic regime attempted to implement a similar transformation in Guatemala in the early fifties, but it ran into the violent opposition of the US-owned banana plantation owners who organized, with the help of the CIA, a military coup in 1954 that introduced almost half a century of violent repression, revolutionary movements and civil wars. In Peru, a nationalist and populist military regime responded to Indian peasant mobilization in the Andean highlands in the late sixties, but the experiment did not last more than a decade, when other military rulers phased it out. In all of these countries, Indian peasantries were at the center of conflict.

The Cuban revolution and its land reform legislation in the early sixties was perceived by the ruling oligarchies of Latin America as highly threatening to their stability because of its influence on the rest of Latin America. In response to this 'Communist' threat, the Kennedy administration in the United States proclaimed the Alliance for Progress, a program designed to mollify peasant demands and to stem the appeal of the Cuban revolution through the enactment of mild land reform legislation and some economic aid to a number of Latin American governments.

The Sandinista revolution of 1979 intended to carry out land reform in Nicaragua, which came to naught a decade later when the Sandinistas were voted out of power. A similar fate befell the Chilean land reform during the years of President Salvador Allende, which was cut short by the military coup of 1973. Sometimes Indian communities did indeed receive some land and benefits, in other instances land reforms simply passed the Indians by. Consequently, access to land has become a major claim of indigenous organizations and the subject of continuous disputes between Indian communities and the State in much of Latin America.

A second feature which definitively marked the situation of indigenous peoples within the State was the non-recognition of Indian cultural and social identity as part of national society. The founding fathers and intellectual elites of the fledgling republics grandly ignored demographics and based the projects of their national societies on their self-perception as a Western, Catholic, racially European people. Indians and Africans were excluded from this project. These ruling groups tried hard to be accepted at the court of Western civilization and to build nations in the image of Western political and economic models. They borrowed their legal systems and public administrations from Spain and France, their political constitutions from the United States, their economic liberalism from Great Britain, their military codes from Prussia. They wished to improve their racial stock in true Darwinian fashion and imported immigrant settlers from Europe. Indians and Africans were considered a burdensome obstacle to nation-building. Wherever it was impossible or too cumbersome to eliminate the latter physically, they were either segregated in the backlands to wither away or remain as an inexhaustible supply of cheap labor, or else they were forced or encouraged to shed their evil cultural ways and become 'nationalized' as it were, that is, to turn into useful citizens of the state according to the hegemonic cultural model.

4.1 Indians, Mestizos and State Policies

Moreover, the *criollo* elites were gradually challenged by the growing *mestizo* population, who came to occupy the ethnic middle ranks and frequently became identified with middle-class political parties and nationalist politics. Indians and their cultures were expected to disappear. Acculturation and *Ladinoization* (becoming Ladino or Mestizo) were seen as inevitable processes, a part of the general tendency towards modernization. The latter alternative became official *indigenista* policy in the twentieth century, when governmental programs stressed assimilation and integration of the indigenous through communications and road building, the market economy, education and community development.

Indigenismo became the domestic expression of assertive nationalism and populism in the twentieth century. It was, during the early decades, a generous, inspiring, progressive ideology. Its proponents, mainly *mestizo* anthropologists, were convinced that they were not only serving their countries well but helping the indigenous overcome their many limitations on the way to becoming modern useful citizens. Directed culture change and applied anthropology were the conceptual tools necessary for this grandiose enterprise: soon our countries would become modern and Indians would be only relics of a picturesque past. Indeed, magnificent museums–such as the one in Mexico City–were built to pay homage to the great dead civilizations of the past and to symbolize the strong Indian roots of the contemporary *mestizo* nation.

As defined by the then prevailing nationalist ideologies which had arisen during the preceding century, the urban *mestizo* middle class intellectuals and their political offshoots usually rejected outright the indigenous components of the national culture and actually saw no future for them, except in an idealized past whose privileged locus turned out to be the museums, and more recently as an instrument for earning foreign exchange in the form of tourism and handicrafts. Indigenismo evolved into a set of social policies intended to 'integrate' the numerous Indian communities and tribes into the life of the nation, as defined by the governing elites. It has always been the instrument of the national state for dealing with the Indians, rather than a process whereby the Indians themselves could determine their own destiny.

Latin America's ruling classes, unable to wish Indians away, have always been quite happy to build nations without Indians. To their discomfort, as the new millennium unfolds, not only are indigenous peoples still there–and their numbers are rising–, but they are actually challenging the very model of the nation-state that the ruling groups have been trying so conscientiously to build up.

Who are the Indians in Latin America, and how many are there? Whereas criteria used in definitions vary from country to country and census data are notoriously unreliable, knowledgeable estimates consider that there are over 400 different identifiable Indian groups, with a total population of close to forty million, that range from some numerically insignificant, almost extinct jungle bands in the Amazon to multi-million strong peasant societies in the Andes. Mexico has the largest indigenous population in Latin America, of between 12 and 15 % of the total population of around 100 million. In Guatemala and Bolivia Indians make up the majority of the national population, and in Peru and Ecuador they come close. Though Brazil's Indians represent less than one half percent, as the original inhabitants of the Amazon basin they have played a significant role in recent years, resisting encroachment upon their territories, claiming land rights and political representation, struggling for the preservation of the Amazon environment and achieving their incorporation into the new Brazilian constitution adopted in 1988.

The rise of indigenous organizations over the last three decades may be considered both as a cause and effect of the transformations occurring in the public sphere regarding indigenous peoples in Latin America. Back in the nineteen sixties there may have been only a handful of formal organizations created and run by indigenous persons and pursuing objectives of interest to indigenous peoples as such. By the middle nineties the turn of the century we are speaking of many hundreds of such associations, of all types and kinds: local level organizations, inter-communal and regional associations, formally structured interest groups, national-level federations, leagues and unions, as well as cross-national alliances and coalitions with well developed international contacts and activities. Indigenous organizations, their leadership, objectives, activities and emerging ideologies, constitute a new kind of social and political movement in contemporary Latin America.

One of the earlier successful organizations is the Shuar Federation, established in the early sixties to protect the interests of the various dispersed Shuar communities in the Amazon lowlands of eastern Ecuador. The Shuar decided to form their federation to defend their land from encroachment by outside settlers and commercial interests, and in the process discovered that the struggle over land rights could not be separated from their survival as an ethnically distinct people with their own traditions and cultural identity.

Similar organizations emerged during the seventies in several other countries, and they consolidated their activities during the nineteen eighties. While often community based, some were able to build larger units, involving an increasing number of local communities and emphasizing ethnic identity as a unifying bond and a mobilizing agent. A number of ethnic organizations began to appear on the political scene, whose leaders would speak in the name of the ethnic group as such rather than for this or that particular rural community. This level of organization was soon followed by region-wide associations, involving several ethnic groups.

Thus arose the *Confederation of Indigenous Nationalities of Amazonic Ecuador* (CONFENAIE), the Indigenous *Association of the Peruvian Jungle (AIDESEP), the Regional Indigenous Council of the Cauca Valle*y (CRIC) in Colombia, the *Indigenous Confederation of Eastern Bolivia* (CIDOB), and many others. They all held conferences and congresses, published manifestos and platforms, addressed petitions to state and national governments, as well as the international community, and often organized militant actions such as protest marches, demonstrations, sit-ins, and land occupations, practiced active resistance, or initiated legal proceedings, lobbied legislatures and public officials, to further their various objectives.

A more recent level of organization became the country-wide indigenous confederation. Again, the *Ecuadorean Confederation of Indigenous Nationalities* (CONAIE) has been in the forefront of political activity, by organizing two major peaceful indigenous 'uprisings' in Ecuador in 1990 and 1993 that practically paralyzed the country and forced the national government to negotiate with the indigenous peoples over agrarian and other issues. Later they organized their own political party and in January 2000 they joined a group of army officers, staged a coup and ousted an elected president before they were forced to withdraw. In 2006 they allied themselves once again with a winning presidential candidate. They had finally arrived in Latin American politics.

The Brazilian *Union of Indian Nations* (UNI), made up of numerous Amazonian tribes, played a crucial role in the political discussions leading up to the drafting of the constitutional article devoted to indigenous peoples in the Brazilian constitution (1988) Similarly the *Organización Indígena de Colombia* (ONIC) took an active part in the national political debates that resulted in Colombia's constitution of 1991, and indigenous movements were actively involved in the drafting of the new Venezuelan constitution of 1999.

Indigenous organizations have also reached out beyond their country's borders to become involved in international activities. Particularly relevant has been their participation in the United Nations Commission on Human Rights (now the Human Rights Council), which adopted a Declaration on the Rights of Indigenous Peoples in 2006. A similar regional declaration is being prepared within the framework of the Organisation of American States. In the meantime, a number of states in the Americas have ratified Convention 169 of the International Labour Organisation that refers to the rights of indigenous peoples.

They regularly attend meetings at the United Nations and have actively supported the establishment of the Permanent Forum on Indigenous Issues and the appointment of a Special Rapporteur on the Human Rights of Indigenous Peoples by the UN Human Rights Commission. They have also successfully sued the government of Nicaragua in the Inter-American Court of Human Rights for a violation of their internationally and constitutionally recognized land rights. This Court as well as the Inter-American Commission on Human Rights have in recent years ruled favorably on the rights of indigenous communities.

A careful analysis of the declarations, resolutions and statements produced by these organizations and congresses would show a progression of ideas and a sequence of issues of concern to their members over time. In earlier years, indigenous manifestos would remind the public at large of their historical victimization and their secular poverty, and demand some kind of overall retribution and justice from governments. At the same time, a persistent theme in many of these documents was a certain idealization of the Indian pre-Columbian past, depicted as a kind of 'Golden Age', a period without exploitation, discrimination and conflict, even as Indian cultures are described as morally superior to Western civilization.

In later years, the demands put forward by indigenous organizations became more focused on specific issues such as land, agricultural credit, education, health, technical aid, investments in infrastructure and so forth, issues that the State is expected to solve. More recently, socio-economic demands have been coupled with calls for autonomy and self-determination. Ethnic identity has become a rallying point for many such organizations, concern over the ecology is now a major issue of contention, particularly in the Amazon lowlands, and increasingly there are demands for legislative changes and compliance with recent international legal instruments. At international meetings indigenous representatives are increasingly putting forward constructive proposals involving new legislation and social and economic policies. The new slogan of some of the indigenous movement is: "From Protest to Proposals".

Who represents the indigenous peoples? The traditional village elders are being displaced by a younger generation of cultural brokers. As more and more indigenous professionals appear, intellectual indigenous elite has emerged in several Latin American countries that is becoming the life-blood of the new organizations. Indigenous intellectuals are actively engaged in developing the 'new indigenous discourse' that gives these organizations their distinctive identities. They are not only involved in formulating the political agendas of their movements, they also rediscover their historical roots, are concerned with language, culture and cosmology, and become actively engaged in 'inventing traditions' and building 'imagined communities'. To the extent that this intelligentsia participates in national and international networks and is able to mobilize material and symbolic resources, indigenous intellectuals have become indispensable links in the process of organization and mobilization.

In Mexico, for example, the first formal indigenous associations beyond the local level were organized by Indian school teachers working for the federal Ministry of Education. They had been trained to teach in bilingual grade schools in Indian villages. A *National Association of Writers in Indigenous Languages* brings together native students of indigenous oral traditions as well as creative writers, most of whom are at the same time employees of the government or active academics. During the tragic years of civil war and military repression in Guatemala, Indian mobilization frequently took place through innocuous-sounding associations for the preservation and study of Maya culture, by which an emerging Maya identity has become crystallized.

On the other hand, the indigenous leadership also draws support from the 'grass roots', from local activists engaged in struggles against human rights abuses, or for land rights, or over environmental concerns, issues in which indigenous women are often especially active. Sometimes there appears to be a tension between the grass-roots activists and the intellectuals, because the former are concerned with more immediate issues and push for concrete solutions, whereas the latter are more involved in institution-building over the long term. Moreover, whereas indigenous intellectuals are contributing to the development of an 'Indianist' ideology and Weltanschauung, and also at times find themselves engaged in discussions with various other ideological tendencies in Latin America (nationalism, Marxism, liberation theology, Christian democracy, evangelical Protestantism), grass-roots activists do not have much patience for intellectual debates and are more interested in negotiating specific issues with 'the powers that be' rather than aspiring to ideological coherence or purity. These various approaches, as well as other factors, have led to not a few disputes over organizational matters, strategy and tactics that sometimes give the impression of a very fragmented and factionalized indigenous movement.

There are a number of factors which may account for the rise of indigenous awareness and the emergence of these new social movements on the public scene. In the first place, mention could be made of the overall disenchantment with, and the failure of, traditional development policies which were assiduously pursued by national governments and multilateral organizations since the end of the WWII. 'Economic development' was the magic slogan wielded by generations of government planners and academics, that would bring improved living standards and burgeoning incomes to the poor, the marginalized, the backward populations of Latin America. This did not happen, as the 'lost decade' of the nineteen-eighties so clearly showed. Indigenous populations were indeed drawn into the 'modern' sector of the economy through market mechanisms, labor migrations, expanding infrastructure in means of communication and transport, but they saw the benefits of growth going, as always, to the elites. A World Bank study reports that the indigenous populations are for the most part poor or extremely poor and that their living standards are 'abysmal'. This fact was not lost on the emerging Indian intelligentsia who soon became skeptical of upbeat economic projections, government promises, and predictions about their imminent accession to progress and civilization. Disillusionment with mainstream development strategies was shared widely, beginning in the nineteen seventies, and the search for 'development alternatives' often focused on the local grassroots level which would naturally include indigenous communities. Things have not changed since then. The Inter-American Development Bank reports that poverty and underdevelopment has increased in Latin America over the last decade, and a study carried out be World Bank indicates that in ten years there has been no significant change of development indicators among indigenous peoples in the region.

A second factor accounting for the rise of indigenous movements was the increasing awareness by the emerging Indian intellectuals that the modern nation-state which the *mestizo* elites had been building so assiduously ever since the nineteenth century was fundamentally flawed. Instead of being all-inclusionary, it was in fact exclusionary: Indian cultures were denied, Indians were victims of subtle or open racism and discrimination; indigenous peoples (even when they constituted demographic majorities as in Bolivia and Guatemala, and in numerous sub-national regions elsewhere), were excluded from economic well-being, social equality, political decision-making processes, and access to justice in the legal system. Indians could not recognize themselves in the prevailing model of the 'national' state, as fashioned by the mestizo and white upper-class elites.

The indigenous roots of Latin America had long been considered a burden by the European elites, and government assimilationist policies made it clear that indigenous cultures had no future in the modern nation-state. Despite formal citizenship indigenous peoples have in fact been treated more frequently as second-class citizens. Many indigenous persons in fact interiorized the stereotypes and stigmas imposed on them by the dominant sectors, and resorted to self-denial and self-denigration in order to become accepted by non-Indians. Others developed a 'culture of resistance', turning inward, avoiding contact as much as possible with the outside world. Still others, realizing that the existing model of the nation-state denies them their identity and their very survival as viable cultures, have begun to challenge the dominant mainstream notion of the nation, by proposing alternative conceptions of a multi-cultural, poly-ethnic state. This is one of the demands that the new indigenous movement has been putting forward in recent years.

The more traditional political parties and institutions of government and the civil society were slow in recognizing the significance of the emerging indigenous movements. During the nineteen eighties, however, a number of processes and tendencies impinged upon public debates. At the international level, cold war ideological confrontation in Latin America came to a virtual end with the break-up of the communist world, although the US still actively pursued it in Cuba, El Salvador and Nicaragua. Secondly, the global economy, which had never been absent from Latin America since colonial times, reaffirmed its impact on the rural areas, including indigenous territories, as in the Amazon basin, Central America, southern Chile and elsewhere, generating tensions and conflicts between Indian peasantries, state institutions and trans-national corporations. Thirdly, a cycle of authoritarian military interventions in politics (which had been linked to the 'national security ideology' of the cold war era) came to an end, and a number of Latin American polities began what has been grandly (and perhaps somewhat overoptimistically) called a democratic transition, liberating the forces of civil society for electoral competition and opening a formerly restricted political space to new or re-emerging social actors.

The struggle for democracy gave rise in Latin America to an articulate and militant human rights movement, which soon became deeply involved in the issues of indigenous peoples. The Inter-American Commission and Court of Human Rights were increasingly besieged by complaints concerning human rights abuses against indigenous people, and relevant UN committees received reports and complaints on the situation of indigenous human rights. It is almost impossible to chronicle the many associations, caucuses, committees, councils, congresses, conferences, symposia, workshops and meetings that activated Indian agency where none or little had existed before. Many such organizations have not survived, others changed over the years, and still others grew and developed true to the stages and cycles of the various theories of social movements.

The new Indian movement in Latin America has not yet produced a specific coherent ideology, and perhaps it has no need for it. But it is developing a new discourse, which has changed the way the wider society sees the Indians and the way they see themselves. Most of all, the movement and its various expressions are changing the relations between indigenous peoples and the State. In this context must be placed the constitutional and legislative changes that were made in the last two decades of the century in a number of the region's states, legally enshrining indigenous rights, in many cases for the first time. To mention only a few: Bolivia, with a majority of Indian population, amended its constitution in 1994 and adopted special laws recognizing that the country is multi-ethnic, multicultural and multi-lingual. In 2005 for the first time Bolivia elected an indigenous president by a large majority, who is pursuing a social and economic agenda geared towards the demands of the indigenous electorate. Brazil's constitution of 1988 devotes an entire new chapter to the Indians. The 1991 constitution of Colombia grants important autonomic rights to its indigenous populations, and in 2007 Ecuador was engaged in a constitutional review along similar lines. The constitutions of Guatemala (1986), Nicaragua (1987) and Venezuela (1999), Panama (1972), Paraguay (1992) and Peru (1993) have no less important constitutional statements. The latest constitutional reform on indigenous rights was adopted by Mexico in 2001.

In other countries, such as Argentina and Chile special legislation concerning Indians was adopted in the post-dictatorship years. While these legal advances are surely important in themselves, the open question is how the new legislation will be implemented and how Indian communities will benefit. The answer is not at all clear, because complaints are increasingly heard that the new laws are not being applied as they should, or that secondary legislation has not been adopted after general principles were laid down in the constitutions.

Students of these constitutional reforms point to several commonalities in the process:

1. The rights of indigenous peoples are recognized in the political constitution, rather than ordinary law or decree, giving them a higher symbolic and juridical rank;
2. In some cases (Colombia, Ecuador, Paraguay, Venezuela) indigenous peoples themselves directly participated in the drafting of some of the new constitutions;
3. The new reforms present a new conception of national identity that is multiethnic and culturally diverse, which replaces a prior homogenizing national myth;
4. The new reforms recognize rights that are collective in nature, and that in some cases grant distinct powers or resources to indigenous communities or populations, as opposed to individuals; and
5. The new reforms restore the colonial tradition of recognizing the public authority and jurisdiction of indigenous authorities (usually at the community

or municipal level) and self-governing structures over some issues, including the exercise of indigenous customary law.

The struggle for indigenous rights is still in its infancy and after the promising beginnings mentioned before, the going will be rough from now on. There are several reasons for this, one being that the opponents to Indian rights have now been able to organize and mount a counter-offensive; another one, that after the first break-through on the political scene, Indians and their allies have not been able to set themselves clear short and medium term objectives, nor were they able to develop an effective political strategy to achieve their aims. This seems to have alienated a number of potential sympathizers in the general population and the political establishments. A case in point was the failed civil-military coup in Ecuador in January 2000, in which a prominent Indian leader played a key role. In Guatemala a referendum on the incorporation of indigenous rights into the constitution, as agreed upon in the 1996 peace agreement which put an end to over three decades of brutal civil war, did not receive majority approval contrary to widely held expectations. While there are increasing numbers of indigenous parliamentarians in many countries, who represent different political parties, there is no clear pattern of ethnic voting nor can any political party count on the automatic contribution of an indigenous electoral bloc. In general, it may be said that indigenous demands are channeled in other ways than through traditional electoral party politics, but this may change in the future.

A crucial issue today is the debate concerning demands for indigenous territorial autonomy, as well as access to, and control over their own natural resources. Some of the recent legislation recognizes these rights on paper, but they are difficult to implement. Litigation over such matters has now been taken up in international human rights institutions. In August 2001 the Inter-American Court of Human Rights found that the government of Nicaragua had violated the property rights of the Awas Tigni Indian community in granting logging rights to a foreign corporation on Indian lands in contravention of Nicaragua's own laws.

The Constitutional Court in Colombia has decided several cases in favor of Indian communities against actions of the government, based on the country's new constitution. The demarcation of recognized Indian lands and territories is a lengthy and often conflictive process, as in Brazil and Panama. The Mapuche of southern Chile are involved in a struggle against the privatization of their traditional territories, decreed during the Pinochet dictatorship.

The meaning of autonomy is ambiguous and its complexities are many. Most of the issues are not resolved in the new legislations, and specialists cannot seem to agree on the details. In fact, most governments in Latin America, permeated by a longstanding centralist tradition of authority are leery of autonomy, especially when related to indigenous peoples. The concept of autonomy and the self-determination of peoples has now become a point of honor for the indigenous movement, and self-determination appears at the top of the list of rights claimed in almost every indigenous political document. Progress on the draft declarations on indigenous rights is currently stalled within the United Nations and the Organization of American States in great measure because of lack of consensus on these points.

The Zapatista rebels in southern Mexico, who staged an armed uprising in January 1994, reached an agreement with the federal government after many months of negotiations in 1996, on which the government later reneged. In August 2001 the national congress approved a constitutional amendment that included some of the elements of the peace agreement, but the Zapatistas and the indigenous movement are dissatisfied with the outcome, claiming that the reform does not comply with the major points of the peace accord. An appeal to the Supreme Court by hundreds of indigenous municipalities was unsuccessful. Further negotiations between the government and the Zapatistas were interrupted and the historic conflict in Chiapas is as yet unresolved.

An equally conflictive issue concerns the controversy over individual versus collective rights. Countering the liberal state's assurances that every human being enjoys a packet of inalienable individual liberties, indigenous rights advocates argue that even the best of fundamental freedoms can hardly be enjoyed by ethnic groups and minorities who are systematically discriminated against and excluded by the power structure and the prevailing system of social stratification. Something more is needed, and this would be a bundle of group rights allowing the indigenous to fully live and reproduce their cultures, organize their lives according to their own social norms, maintain and develop their own collective identities, enjoy social, political and legal status as distinct groups in the wider society, and relate to this society and the national state on their own terms as recognized and respected peoples or nations.

No doubt the recognition of these collective rights requires a complete overhaul of the national state, of this 'imagined community', the nation, which the *criollo* and *mestizo* elites created to serve their own interests. Arguably, individual human rights cannot be fully enjoyed by members of discriminated against subaltern groups, unless these are acknowledged as equal and full partners in all their distinctiveness and dignity within a nation-state. Thus the recognition of group rights may be seen as a condition for the enjoyment of individual rights, but they are not easily acknowledged in Latin America's legal systems.

4.2 Ethnic Cultures versus National Culture?

Behind many of the controversial issues over which indigenous peoples and the State in Latin America square off, none has raised more polemics than indigenous cultures and identities. The almost bi-centenary old idea of a single national culture has been put to a severe test by indigenous demands for bilingual and intercultural education and by the relatively recent legal recognition in some states that these countries are pluri-ethnic and multicultural. The current debate in Guatemala expresses these conflicting views rather well.

During the terrible years of the civil war, in which hundreds of thousands of indigenous people were murdered by the army, and many more became refugees, one of the few spaces of resistance of the Maya population were local cultural associations, which grew in numbers and activities after the middle eighties when the military ceded formal power to elected civilian governments. Indigenous intellectuals developed a new discourse of Maya cultural identity, which was strengthened by the signing in 1996 of the peace accord between the government and the guerrillas, the major agreements being the one on indigenous rights and culture. The Pan-Maya cultural movement spread rapidly and has contributed to changing both official discourse and the demands of political and social organizations. In a *criollo* and *ladino* dominated state, the majority Maya people have always been considered outsiders, and were effectively excluded qua Indians from the society and the polity. The civil war and the ensuing peace accord have changed all that. The various indigenous ethnic groups are now coalescing into a newly constructed Maya identity (including the revival of Maya religion—this in a traditional Catholic country in which Protestantism has made considerable inroads in recent years). Maya intellectuals and activists see themselves as opposing the hegemonic mestizo 'national' identity, and claim for their people not only a major cultural role in the redefinition of the nation, but also political representation and access to power.

The Maya cultural movement has developed various theoretical and policy perspectives, and it speaks through different, sometimes dissonant, voices. For example, there is no agreement as to whether the Maya people are to be considered as only one nationality or many. Demerio Cojtí (a Kakchikel Indian who is vice-minister for multicultural education) speaks of 20 Maya nationalities in Guatemala alone (there are others in neighboring countries). Should the new politico-administrative divisions in the country be based solely on Maya ethnic identities, or also include Ladinos? Should political representation in congress reflect exactly the ethnic make-up of the country? How many of the Maya tongues should be recognized as official languages, and in what way shall multi-lingual and inter-cultural education be implemented in the school system?

The search for and the construction of a new Maya identity in Guatemala does not enjoy universal approval. The Maya culturalist positions have been attacked, among others, by some who would like to see ethnic identities subsumed in a wider class-based popular alliance, and those who argue from a Ladino perspective that Maya 'essentialism' is no more than an artificial construct promoted by a host of internationally financed non governmental organizations. According to this view, there is no such thing as a Maya nation or people, and Maya activists are accused of becoming anti-Ladino racists in turn, the only valid solution to Guatemala's problems being the development of an intercultural *mestizaje* in which Indians and Ladinos would learn to co-exist and interact on equal terms.

Half a century ago, when states in Latin America began to carry out policies for the development of indigenous communities, Indian populations lived mainly in isolated rural villages. Ten years ago, indigenous organizations complained that the celebrations the Western world was organizing around the quincentenary of the misnamed 'discovery of America' was for them a time of mourning and remembrance. Today, indigenous representatives sit on United Nations bodies, in national parliaments and government cabinets. An emerging corpus of indigenous law is expressed in international legal instruments and national legislation. Indigenous peoples claim recognition and a distinct place in plural and multicultural societies. The traditional concept of the homogeneous national state has been challenged by the 'return of the natives'. Far from disappearing in a *mestizo* melting pot, over 400 indigenous groups in Latin America demand attention to their rights and problems, an end to racism, discrimination, poverty and social exclusion. Indigenous peoples claim lost lands and territories, respect for languages and cultures, the right to practice their laws and customs, and a measure of autonomy within the territorial state. Above all else, they want to be accorded a minimum of human dignity, which for so many centuries has been denied to them. Surely this is not too much to expect after half a millennium.

Rodolfo Stavenhagen with Masai women in Kenya. *Source* Personal photo collection of the author

Chapter 5
Indigenous Peoples in Comparative Perspective (2004)

Abstract Indigenous Peoples in Comparative Perspective—Published as Occasional Paper 2004/14 by the United Nations Development Program (UNDP), this was a background paper for the United Nations Human Development Report 2004 devoted to Cultural Liberty in Today's Diverse World. (This document is in the public domain).

5.1 Who and Where

One of the more widely debated topics involving indigenous peoples and state policies is the ambiguity surrounding the term 'indigenous'. There is no internationally agreed upon definition of indigenous peoples. Different states adopt—if at all—definitions in terms of their particular contexts and circumstances. The term indigenous is frequently used interchangeably with others, such as 'aboriginal', 'native', 'original', 'first nations' or else 'tribal' or other similar concepts. In some countries no formal designation exists even though there might be general agreement that such populations do in fact inhabit certain areas of the country. In others, the existence of indigenous groups is denied altogether and therefore their definition becomes even more problematic. Yet the absence of an international definition does not prevent constructive action in the protection of the rights of indigenous peoples and the promotion of their development. Whereas at the national level, formal definitions have become more common in recent years, the criteria of membership in an indigenous group, nation or community are not always clearly established.

While indigenous peoples the world over share many commonalities, there are also great differences among them. In North America, for example, Native Americans were considered as sovereign and distinct by the governments of the

R. Stavenhagen, *The Emergence of Indigenous Peoples*, Texts and Protocols 3,
DOI: 10.1007/978-3-642-34144-1_5, © The Author(s) 2013

United States and Canada well into the nineteenth century. Relations between these peoples and the state were based on treaty arrangements which, as time went by, were abrogated unilaterally by the states concerned; a similar process occurred in Chile. In New Zealand, on the other hand, the Treaty of Waitangi sets the frame of the relations between the Maori and the State. In all these countries, the situation of indigenous peoples today results in many ways from such agreements and their consequences.

According to recent estimates, the indigenous populations number around 300 million, most of them in South Asia. In Latin America the number of Indians is estimated at around 40 million people, which means roughly 10 % of the region's total population, and they are concentrated mainly in the Andean area, Mexico and Central America, comprising about 400 different groups who are in turn divided into many thousands of local communities.

Of special interest is the situation in Africa, where states usually do not recognize indigenous peoples among their populations because of their fear of resurgent 'tribalism' and because most Africans consider themselves to be 'indigenous' to the continent. The African Commission on Human and Peoples' Rights now includes an expert working group on the rights of indigenous or ethnic communities. At its 2003 session the working group presented a report that "calls for a recognition of the unique character of indigenous people and develop policies and practices in consultation with the people concerned and with due regard to [their] identity".

The countries of Asia also present different approaches. Some scholars consider the various 'tribal' categories used in some states as equivalent to the concept of indigenous peoples but others disagree. The Adivasis in India might usually be considered as indigenous peoples, but they are not officially recognized as such. Similarly the Ainu of Japan have only recently been recognized as a minority or an indigenous people by government authorities. No such definitional problems arise concerning indigenous peoples in Australia, New Zealand, Malaysia and Philippines, nor in Russia where a longstanding legal tradition provides clear guidelines regarding definition of groups and membership criteria. In northern Europe the Sami are also recognized as indigenous.

A United Nations report provides a widely used definition,[1] to which Convention 169 on Indigenous and Tribal Populations of the International Labor

[1] "Indigenous communities, peoples and nations are those which, having a historical continuity with preinvasion and pre-colonial societies that developed on their territories, consider themselves distinct from other sectors of the societies now prevailing in those territories, or parts of them. They form at present non-dominant sectors of society and are determined to preserve, develop and transmit to future generations their ancestral territories, and their ethnic identity, as the basis of their continued existence as peoples, in accordance with their own cultural patterns, social institutions and legal systems." (EC/N.4/Sub.2/1986/7/Add.4, p. 379).

Organization adds: "Self-identification as indigenous or tribal shall be regarded as a fundamental criterion…"[2]

The *Draft Declaration on the Rights of Indigenous Peoples, being considered by the UN Commission on Human Rights*, states the right to membership in an indigenous community.[3] (See footnote 3) The UN Working Group on Indigenous Populations proposes four principles to be taken into account in any possible definition of indigenous peoples: (a) priority in time, with respect to the occupation and use of a specific territory; (b) the voluntary perpetuation of cultural distinctiveness, which may include the aspects of language, social organisation, religion and spiritual values, modes of production, laws and institutions; (c) self-identification, as well as recognition by other groups, or by State authorities, as a distinct collectivity; and (d) an experience of subjugation, marginalization, dispossession, exclusion or discrimination, whether or not these conditions persist.[4]

The lack of adequate legal provisions has led to human rights violations of indigenous peoples in many parts of the world. Ambiguities in their legal situation are of particular concern to indigenous peoples in several Latin American countries, such as Argentina, Chile and Mexico. The Amazigh (also known as Touareg, Kabyles and Berbers) of North Africa demand legal recognition and the respect of their cultural and social rights as an indigenous people. In the Philippines, the provisions of the Indigenous Peoples' Rights Act are sometimes overridden by other legislation, creating tensions and judicial controversy. Within the framework of a constitutional review in Kenya, the Ogiek, a hunter-gatherer people, claim recognition as a distinct indigenous minority. The Masai pastoralists are considered to be an indigenous minority in several east African countries, and their legal recognition varies in the different states. In Malaysia, special legislation applies to the Orang Asli as an aboriginal people. The UN Committee on the Elimination of Racial Discrimination has expressed its concern for the legal status of the indigenous peoples of Cambodia, particularly as regards their rights, culture and traditional lands.[5]

---

[2] Convention 169 of the ILO applies to: "peoples in independent countries who are regarded as indigenous on account of their descent from the populations which inhabited the country, or a geographical region to which the country belongs, at the time of conquest or colonization or the establishment of present state boundaries and who, irrespective of their legal status, retain some or all of their own social, economic, cultural and political institutions.".

[3] "Indigenous peoples and individuals have the right to belong to an indigenous community or nation, in accordance with the traditions and customs of the community or nation concerned." (Art. 9) and "Indigenous peoples have the collective right to determine their own citizenship in accordance with their customs and traditions." (Art. 32).

[4] UN Doc. E/CN.4/Sub.2/AC.4/1996/2. A similar approach is followed by the Indigenous Peoples of Africa Coordinating Committee (IPACC). See IPACC, Annual Report (November 1998 to October 1999) Appendix, 22.

[5] CERD/C/304/Add. 54.

5.2 Background

Ever since the Europeans first came to the shores of the American continent to conquer and settle, indigenous peoples have suffered discrimination, exploitation and racism. During 300 years of colonial domination, the indigenous societies of Latin America were subjected to the worst forms of oppression and exploitation. Much of the colonial wealth of Europe was based on the use Indian workers in the mines and in the fields, which next to African slavery provided an inexhaustible supply of servile labor. Indigenous cultures were destroyed or subordinated to the dominant Hispanic Catholic mold. In some parts of the continent, widespread genocide of indigenous societies took place. The indigenous population, which was decimated as a result of military conquest, ecological destruction, forced labor and the introduction of lethal diseases brought by the colonists, decreased drastically as a result of the invasion. After political independence in the nineteenth century, Indians were formally granted citizenship in most of the new states, though in many parts they remained less than equal to the population of European or mixed ancestry.

Expanding agrarian capitalism destroyed numerous surviving indigenous communities. As in the United States, in Argentina and Chile the still existing sovereign Indian nations were subdued by the national state. Despite holding legal citizenship rights, they were in fact excluded from equal participation in the economic, social and political system. Special legislation often placed indigenous populations at a disadvantage in relation to the rest of society, even when some laws were of a protective and tutelary nature, and in some countries Indians were still treated as minors and legally incompetent until very recently. It was not until the last decades of the twentieth century that most Latin American countries carried out a spate of constitutional reforms regarding native languages and cultures, the recognition of indigenous communities and their territories as specific forms of social organization, as well as the aims of public policies aimed at indigenous populations.

5.3 The Land Issue

For most indigenous peoples survival is the major challenge in a world that has systematically denied them the right to existence as such. Historically linked to the land as the source of their main livelihood, the indigenous have long struggled to gain and keep access to this precious resource which is at the same time the essential element of their identity as distinct cultures and societies. Land rights are the major issue faced by native peoples around the world and they are at the center of numerous conflicts involving indigenous communities, particularly as a result of globalization. The impact of new economic processes can be dramatic, as seen in agricultural modernization, for example. The wide-spread introduction of commercial crops for export, based on the intensive use of modern inputs (mechanization, improved grains, fertilizers, insecticides, and more recently, genetically modified seeds) tends to displace traditional subsistence farming, on which most indigenous communities depend for their survival. Increasing production costs and the need for economies of scale favor the consolidation of larger productive units and integrated agribusiness, putting traditional farms at a disadvantage in highly competitive markets. Agricultural development policies, instead of helping small subsistence farmers overcome their handicaps, have in fact pushed the poorer peasants out of business and favored the concentration of larger agro-industrial enterprises, and they have forced the peasants to become increasingly dependent on, and therefore vulnerable to, the globalized agricultural economy. Current negotiations concerning agriculture within the framework of the World Trade Organization do not bode well for the continued existence of indigenous farming.

From time immemorial indigenous peoples maintain a special relationship with the land, their source of livelihood and sustenance and the basis of their very existence as identifiable territorial communities. The right to own, occupy and use land collectively is inherent in the self-conception of indigenous peoples and generally this right is vested in the local community, the tribe, the indigenous nation or group. For productive purposes it may be divided into plots and used individually or on a family basis, yet much of it is regularly restricted for community use only (forests, pastures, fisheries etc.), and the social and moral ownership belongs to the community. While such rights are protected by legislation in some countries, powerful economic interests often succeed in turning communal possession into private property. From southern Chile to the Amazon basin to Canada's northern forests; from the tropical jungles of southeast Asia to the bush of southern Africa, there is no longer any territory which is not coveted by some international corporation, either for its mineral wealth, its oil deposits, its pastures, tropical or hard-wood forests, its medicinal plants or its suitability for commercial plantations, its hydraulic resources or its tourist potential. Indigenous peoples are the most recent victims of globalized development, and if these tendencies continue unabated, their chances of survival are becoming weaker, their very existence as distinct societies and cultures is seriously endangered.

Closely linked to the land problem is the territorial issue. Indigenous peoples have been historically rooted in specific locations, their original homelands, which in some cases constitute well defined geographical areas. Indigenous peoples' organizations now demand the recognition and demarcation of these territories as a necessary step to ensure their social, economic and cultural survival. The territory of the San Blas Kuna is constitutionally protected in Panama; so is that of the Yanomami in northern Brazil. The Mapuche of southern Chile and the Miskitos of Nicaragua, among many others, have been in the forefront of these struggles in their countries. The Colombian constitution of 1991 recognizes the traditional homelands of a number of indigenous groups and assures them of legal protection. Philippine legislation recognizes indigenous ancestral domains. In some Canadian provinces aboriginal title to territory is legally recognized.

Convention 169 of the International Labor Organization, adopted in 1989, calls upon States to respect indigenous lands and territories, and proclaims the right of indigenous peoples to control their natural resources. This is a most important right, because many of the current conflicts over land and territory relate to the possession, control, exploitation and use of natural resources. In a number of countries it is the State which keeps for itself the right to control such resources, and in numerous instances multinational corporations are asserting their own economic interests over them, unleashing complicated conflicts over ownership and use-rights with indigenous communities. In Chile, for example, one law recognizes de rights of indigenous communities to their lands, but other laws allow any private party to claim possession of subsoil and water resources on them. Under these circumstances, indigenous communities are hard put to defend their ancestral claims.

Indigenous peoples in Southeast Asia face the loss of control over land and resources due to the non-recognition of customary land rights. In most Southeast Asian states there are no legal rules granting indigenous peoples the right to their land and many indigenous peoples are threatened by logging, mining and other exploitative activities or due to infrastructure programs (dams, roads) pursued by national governments. In Resolution 55/95 on Cambodia, the UN General Assembly notes that illicit logging "has seriously threatened full enjoyment of economic, social and cultural rights by many Cambodians, including indigenous people".[6] A major recent development in this country is the 2001 land law, which states that ownership of land 'is granted by the State to the indigenous communities as collective ownership. This collective ownership includes all of the rights and protections of ownership as are enjoyed by private owners'.

5.4 Development and Indigenous Peoples

Indigenous peoples live mainly in rural environments. Wherever they have been able to maintain their community lifestyles and their traditional cultures, it is because the areas in which they live have been spared major upheavals resulting from rapid economic and ecological transformations. But this situation has changed extensively over the last few decades, as national governments, large corporations and multilateral financing agencies turn their attention to so-called undeveloped regions in order to extract natural resources, establish plantations and industrial plants, develop tourist activities, ports, communication hubs or urban centers, build transportation networks, multipurpose dams, military bases or toxic waste dumps. Wherever such developments occur in areas occupied by indigenous peoples it is likely that their communities will undergo profound social and economic changes that are frequently not well understood, much less foreseen, by the

[6] See A/RES/55/95 of 28 February 2001.

authorities in charge of promoting them. Large-scale development projects will inevitably affect the conditions of living of indigenous peoples. Sometimes the impact will be beneficial, very often it is devastating, but it is never negligible.

Traditionally few governments have taken the rights and interests of indigenous peoples into account when making plans for major development projects. As the projects mature, which may take several years depending on their characteristics, the concerns of indigenous peoples, who are seldom consulted on the matter, take a back seat to an overriding 'national interest', or to market-driven business objectives aiming at developing new economic activities, maximizing productivity and profits. For a long time, multilateral financing agencies involved in the planning and execution of such projects appeared to go along with this approach. Hence, the social and environmental concerns expressed by many people, including indigenous communities, have not been given the necessary attention.

In recent years, this situation is changing, as multilateral agencies, national governments and private companies take up a new interest in indigenous concerns. At the international level, ILO's *Indigenous and Tribal Peoples Convention 169* stipulates that:

1. "The peoples concerned shall have the right to decide their own priorities for the process of development as it affects their lives, beliefs, institutions and spiritual well-being and the lands they occupy or otherwise use, and to exercise control, to the extent possible, over their own economic, social and cultural development. In addition, they shall participate in the formulation, implementation and evaluation of plans and programs for national and regional development which may affect them directly..."
2. Governments shall ensure that, whenever appropriate, studies are carried out, in co-operation with the peoples concerned, to assess the social, spiritual, cultural and environmental impact on them of planned development activities. The results of these studies shall be considered as fundamental criteria for the implementation of these activities.
3. "Governments shall take measures, in co-operation with the peoples concerned, to protect and preserve the environment of the territories they inhabit." (Art. 7)

Numerous international conferences have reaffirmed such rights in one formulation or the other, notably the Rio Earth Summit (1992) and the Johannesburg World Summit on Sustainable Development (2002). The World Bank is in the process of adopting a new operational policy that establishes the need to involve indigenous peoples in development projects that may affect them, and the Inter-American Development Bank has laid down similar guidelines for its own activities. Several states have likewise adopted legislation in the same sense.

None have been more concerned with these important issues than indigenous peoples themselves. One recent study reports on "the disproportionate impacts that indigenous peoples suffer from development programs, so long as their human rights are not fully recognized, and so long as they continue to be marginalized in decision-making affecting their lives".[7] Further, indigenous peoples argue that "as the pressures on the Earth's resources intensify, indigenous peoples bear disproportionate costs of resource intensive and resource-extractive industries and activities such as mining, oil and gas development, large dams and other infrastructure projects, logging and plantations, bioprospecting, industrial fishing and farming, and also eco-tourism and imposed conservation projects."[8] On the specific issue of large dam construction the World Commission on Dams finds that:

"Large dams have had serious impacts on the lives, livelihoods, cultures and spiritual existence of indigenous and tribal peoples. Due to neglect and lack of capacity to secure justice because of structural inequities, cultural dissonance, discrimination and economic and political marginalization, indigenous and tribal peoples have suffered disproportionately from the negative impacts of large dams, while often being excluded from sharing in the benefits."[9]

To the extent that many of these projects are located on the ancestral territories of indigenous peoples, it is not surprising that they should raise the issue of the rights to land, to prior consent concerning its use, to participation in the decision-making process regarding the implementation of such projects, and beyond this, the right to share in the potential benefits. At the twentieth session of the *Working Group on Indigenous Populations* (WGIP) "... virtually every indigenous participant stated that their right to self-determination is a pre-condition for the realization of all other human rights, and must be considered as the bedrock that ensures their self-governance, whereby they can participate in decision-making processes in policies that directly affect them. They therefore reiterated the intrinsic link of the right to self-determination to various other indigenous human rights issues such as the right to land and natural resources, the preservation of cultural identity, and the rights to language and education."[10]

The right to free, informed and prior consent by indigenous peoples continues to be of crucial concern, inasmuch as too many major decisions concerning large-scale development projects in indigenous territories do not comply with this stipulation, clearly set out in para. 6 of ILO's convention 169, which provides that governments shall

---

[7] Jocelyn Carino, 2001: *Overview paper presented to the Workshop on Indigenous Peoples, Private Sector Natural Resource, Energy and Mining Companies and Human Rights* (New York: United Nations, OHCHR): 4.

[8] CSD Indigenous Peoples' Caucus, "Dialogue Paper by Indigenous Peoples," in Indigenous Affairs 4/01, IWGIA, p.14.

[9] World Commission on Dams, 2000: *Dams and Development. A New Framework for Decision-Making. The Report of the World Commission on Dams*. See Chap. 4, "People and Large Dams-Social Performance", particularly the section on Indigenous Peoples. Available at: <www.dams.org>.

[10] Report of the 20th session of the WGIP, E/CN.4/Sub.2/2002/24, Para. 26.

(a) "consult the peoples concerned, through appropriate procedures and in particular through their representative institutions, whenever consideration is being given to legislative or administrative measures which may affect them directly;
(b) "establish means by which these peoples can freely participate, to at least the same extent as other sectors of the population, at all levels of decision-making in elective institutions and administrative and other bodies responsible for policies and programs which concern them..."

Likewise, Article 30 of the UN Draft Declaration on the rights of indigenous peoples also provides that states shall obtain free and informed consent prior to the approval of any project affecting their lands, territories and other resources, particularly in connection with the development, utilization or exploitation of mineral, water or other resources.[11] The Proposed American Declaration on the Rights of Indigenous Peoples (Article 21 [2]) contains a similar provision. For indigenous peoples the principle of free, prior and informed consent is an issue of primary importance. In some states legislation has progressed in this direction. The Aboriginal Land Rights (Northern Territory) Act 1976 (amended in 1987) of Australia not only recognizes the right of Aborigines to own the land, but also provides in effect the right to veto over mining for a 5 year period. Furthermore, a mining grant or a road construction may not be undertaken unless the traditional owners of the land understand the nature and purpose of the proposed mining or road construction proposals as a group and consent to them.[12]

The Indigenous Peoples Rights Act 1997 of the Philippines recognizes the indigenous right to ancestral domain and the land title to traditional lands. Philippine law also requires a developer or company to obtain free, prior and informed consent of indigenous peoples for certain activities, such as (a) exploration, development and use of natural resources; (b) research-bioprospecting; (c) displacement and relocation; (d) archaeological explorations; (e) community-based forest management; and (f) entry of the military. However, indigenous organizations complain that these provisions are not being complied with as they should.[13]

Regarding the exploitation of natural resources in traditional territories of indigenous peoples, the Constitutional Court of Colombia argues that "...indigenous peoples are subjects of fundamental rights. If the State does not guarantee their right to subsistence (survival), these communities will not be able to materialize their right to cultural, social and economic integrity which is stated in the Constitution".[14] Article 2 of the Constitution of Mexico (amended in 2001) recognizes the land rights of indigenous communities but subjects them to the rights of 'third parties, a legal limitation which indigenous organizations and legal

---

[11] See Article 20 of the draft United Nations Declaration on the Rights of Indigenous Peoples.

[12] Paul Kaufmann, Wik, Mining and Aborigines, pp. 15–16, Allen and Unwin.

[13] Jocelyn Carino, pp. 11 –12.

[14] Decision T-652-98 brought by the Embera Katio people with regard to the Urra dam projects.

scholars consider rather as a step backwards in the recognition of their collective rights.

In various UN and other forums, indigenous organizations have signaled their concern about negative impacts of major development projects on their environments, livelihoods, lifestyles and survival. Complaints about loss of land and the lack of control over their natural resources have become widespread. Often these projects entail involuntary displacements and resettlement of indigenous communities that happen to lie in the way of a dam, an airport, a game reserve, a tourist resort, a mining operation, a pipeline or a major highway etc. As a result, violations of civil and political, economic, social and cultural rights occur with increasing frequency, prompting indigenous peoples to launch major protest or resistance campaigns in order to bring public attention to their plight, besides engaging the judicial system or appealing for administrative redress, as well as lobbying the political system. Cases in point:

- The High Court of Australia delivered a devastating decision in 2002, which denied native title rights over any mineral or petroleum resources in the Miriuwung-Gajerrong native title claim first lodged in 1994.[15]
- A number of Pehuenche Mapuche families of Chile have been evicted from the construction site of the Ralco hydro-electric installation which will flood their riverine communities.[16]
- The Sardar Sarovar dam on the Narmada river in India is expected to displace over 320,000 people-among them numerous Adivasi farmers—in what one author has labeled "India's greatest planned human and environmental disaster".[17]
- The San Roque Multipurpose Project in the Philippine Cordillera region is expected to affect several thousand indigenous households, who have protested against the project, but to little avail.[18]
- An indigenous community in Kenya reported to the UNWGIP that "today, this destruction of our cultures and land continues, due to so-called development projects such as mining, logging, oil exploration, privatization of our territories, and tourism."[19]
- It was reported that in Ecuador oil activities are being undertaken which result in the break-up of the traditional, cultural and political structures of indigenous

[15] "No Native Title Over Minerals Or Petroleum," <www.smh.com.au/articles/2002/08/08/0028057983022.html>.

[16] The author visited the site in July 2003.

[17] Cf. Rodolfo Stavenhagen, Second Annual Report to the Human Rights Commission of the Special Rapporteur on the Human Rights and Fundamental Freedoms of Indigenous People, E/CN.4/2003/90, para. 44–52.

[18] Report presented to the author by Äpit Tako, Alliance of Peasants in the Cordillera Homeland and Tebtebba, the Indigenous Peoples' International Centre for Policy Research and Education, October 2002.

[19] Statement of Loodoariak Community Land and Development Programme in Kenya to the 19th session of the WGIP.

communities while facilitating the integration or assimilation of the oil economy in the country.[20]

- In Japan, the building of a hydroelectric power dam in Nibutani, land sacred to the Ainu people, caused the destruction of traditional agriculture and the submergence of their sacred ceremonial sites. It further disrupted the links between the elders and the young as poverty forced families to sell their lands to the Government, which created divisions in the community.[21]
- The Bakun Dam in Malaysia is reported to cause the forced displacement of 5,000–8,000 indigenous persons from 15 communities by clear-cutting 80,000 hectares of rainforests.[22]
- Thousands of families of the Santhal Adivasi people in the Jharkhand province of India have reportedly been displaced as a result of the extraction of the minerals without proper compensation or economic security.[23]
- In Thailand, several highland communities including the Karen people have reportedly been moved out of national parks against their will,[24] whereas tourist development in Hawaii resulted in the displacement of indigenous people and their increasing poverty.[25] Asian indigenous representatives expressed to the Working Group on Indigenous Populations that "...conflict and development interventions had resulted in large-scale displacements, internal and external, and serious consequences for [indigenous] children and youth resulted from the implementation of inappropriate and non—consultative development projects."[26]

African indigenous peoples are not the exception when it comes to displacement from their traditionally owned lands. The creation of national parks or game reserves has forced people off their land. The Boran of Kenya, for instance, testified that four reserves created in Isiolo had been annexed affecting important grazing and watering points previously used by pastoralists. Moreover, the Keiyo indigenous people in Kenya also reported that they have been forcibly evicted from their land without compensation, because of mining activity there.[27] Despite

[20] Report of the Workshop on "Indigenous Peoples, Private Sector Natural Resources, Energy and Mining Companies and Human Rights," E/CN.4/Sub.2/AC.4/2002/3, Para. 104, p. 23.

[21] Report presented to the author during his visit to the site in November 2002.

[22] CSD Indigenous Peoples' Caucus, "Dialogue Paper by Indigenous Peoples," in Indigenous Affairs 4/01, IWGIA: 15–16.

[23] Statement of the representative of the Santhal Advasi People to the 19th session of the WGIP.

[24] Chumpol Maniratanavongsiri, "The Karen Response to Thai Conservation Policies," in CSD Indigenous Peoples' Caucus, "Dialogue Paper by Indigenous Peoples," in Indigenous Affairs 4/01, IWGIA, p.60.

[25] Report of the 19th session of the WGIP, Para. 60–61.

[26] Report of the 18th session of the Working Group on Indigenous Populations (E/CN.4/Sub.2/2000/24), Para.

[27] Statement of the Keiyo Indigenous Peoples of Kenya to the 19th session of the WGIP.

judicial appeal to the country's High Court (which was dismissed on technical grounds) and international concern, the Basarwa (Bushmen) in Botswana had their water supply cut off and have no choice but to leave their traditional hunting grounds for resettlement villages, to make way for government—sponsored development activities in the Central Kalahari Game Reserve.[28]

Evictions or involuntary displacements are so common a feature of major development projects, that the UN Committee on Economic, Social and Cultural Rights comments that forced evictions are prima facie incompatible with the requirements of the International Covenant on Economic, Social and Cultural Rights.[29] Conflicts over development projects on the lands of indigenous peoples lead to further violations of human rights. For instance, forced evictions from their traditional lands may lead to breaches of civil and political rights such as the right to life, the right to security of the person, the right to non-interference with privacy, family and home, and the right to the peaceful enjoyment of possessions. Protesters are often arrested and harassed by the police or private company guards.[30]

For example, people in Penan (Malaysia) have reportedly been arrested because they were blockading roads trying to stop loggers destroying their traditional forests.[31] Philippine indigenous peoples have allegedly been physically abused and detained by mining companies and the police in the process of peaceful picketing against mining activities on their traditional lands.[32] Sometimes, the strict enforcement of environmental conservation laws prevents indigenous farmers, hunters, fishermen or gatherers from using their traditional land or resources, thus turning them into offenders who may be jailed for attempting to subsist.[33] According to a recent report, oil workers in the Upper Pakiria River region of southeastern Peru forced the Kugapakori to move deep into the Amazon and threatened to arrest and decimate the community with diseases if they refused to

[28] Press Release III of the negotiating team, the mandated representatives of the residents of the Central Kalahari Game Reserve (CKGR), April 2002. See also Note No. 01/02/Gen/E/15 II (38) G2 of the Permanent Mission of Botswana in Geneva. The author visited one such village on a personal inspection tour.

[29] Committee on Economic, Social and Cultural Rights, General Comment No. 7, The right to adequate housing (art. 11(1) of the Covenant): forced evictions.

[30] Indigenous communications to the WGIP and the Special Rapporteur on the Rights of Indigenous People; personal accounts provided to the author in Chile, Mexico, Philippines, Canada and elsewhere.

[31] Statement by Tana Uma Amee to the 19th session of the WGIP.

[32] Report of the Workshop on "Indigenous Peoples, Private Sector Natural Resource, Energy and Mining Companies and Human Rights," E/CN.4/Sub.2/AC.4/2002/3, Para. 60, p.15. Similar incidents were reported directly to the author during a visit to the Philippines in December 2002. (E/CN.4/2003/XX/Add.3).

[33] Cf. Robert K. Hitchcock, « Decentralization, Natural Resource Management and Community-Based Conservation Institutions in Southern Africa » p. 39, Indigenous Affairs 4/01, IWGIA; also, personal accounts provided to the author in inter alia Botswana, Canada, Chile, Mexico and Philippines.

leave their home.[34] The Cucapa people in northern Mexico have been restrained by the authorities from practicing their subsistence fishing because of environmental concerns, and the National Commission of Human Rights found that their human rights were being violated and recommended that the Cucapa become participants in the planning and execution of programs for their own social development, including the fishing of protected species for their subsistence.[35]

Major development projects often entail serious health hazards for indigenous peoples. Environmental degradation, toxic chemical and mineral wastes, the destruction of self-sustaining eco-systems, the application of chemical fertilizers and pesticides are but some of the factors that seriously threaten the health of indigenous peoples in so-called "development zones". When relatively isolated indigenous communities enter into contact with the expanding national society and monetary economy—as has happened dramatically in the Amazon basin and other inter-tropical areas in recent decades—indigenous peoples also risk contracting contagious diseases, such as smallpox, aids and venereal diseases, as well as psychological troubles.[36]

Indigenous peoples also argue that "environmental degradation and pollution [are] an integral facet of the health and well-being of indigenous peoples," citing, for instance, toxic contamination by *persistent organic pollutants* (POPs) and other industrially produced toxins.[37] The Batwa of Rwanda report that deforestation leads to loss of traditional medicinal plants and to increased mortality,[38] a complaint also voiced in Canada, Chile, Mexico and Philippines.[39] The construction of a dam in the Cuene region in Namibia will reportedly flood the palm nuts and other trees on which the goats, that provide a vital food source for the riverine Epupa community, feed.[40] Because of the pollution of their traditional lands, the peoples of the north in Russia report that they have now become 'ecological refugees', whereas mining activities in Peru reportedly cause the pollution of fresh water used

---

[34] Cultural Survival Indigenous News, Headlines and Spotlights, 9 August 2002.

[35] Comision Nacional de los Derechos Humanos, Recomendacion num. 08/2002, 30 April 2002.

[36] For reports on Australian aborigines see The second annual report of the Aboriginal and Torres Strait Islander Social Justice Commissioner of Australia, 1994, p.106. On the relation between uranium mining and the health of local Adivasi children in Jharkhand, India, see "The Miner's canary: Indigenous Peoples and Sustainable development in the Commonwealth," Commonwealth Policy Studies Unit, p.32.

[37] Report of the 18th session of the Working Group on Indigenous Populations (E/CN.4/Sub.2/2000/24), para. 29–30.

[38] Ibid., para 31.

[39] Personal communications during in loco fact finding missions.

[40] The Proposed Construction of a Hydropower Scheme on the Lower Cuene River, Indigenous Affairs 2/98, IWGIA, p. 6. It is estimated that "the inundation of the Cuene basin at Epupa would destroy the riverine forests. It would result in loss of an annual crop of hundreds of tons of the palm nuts and would in addition bring an end to gardening in the fertile soils along the riverbank.".

by indigenous peoples for food production.[41] Philippine indigenous representatives reported similar environmental, economic and social effects of mining activities in various parts of the country, which they aptly label 'development aggression'.

Indigenous peoples have argued at length and legitimately that major development projects that do not take into account their fundamental interests entail violations of their basic human rights. At the UN Working Group on Indigenous Populations they maintain that "the indigenous approach to self-development [is] based on the principles of respect for and preservation of land, natural resources and all elements of the natural environment; consensus in decision-making; mutual respect for peoples' values and ideology, including sovereignty over land, resources and the environment under natural law." They also complain that full, meaningful and effective participation of indigenous peoples in development is generally not being considered. For instance, indigenous peoples from Chittagong Hill Tracts in Bangladesh said that "development strategies based on road construction, pacification programs and socio-economic development programs, and immigration, remained in the hands of the military and the participation of indigenous peoples in the development was excluded." The Ogiek of Kenya and the Batwa of Rwanda, referring to the need to get their views across, spoke of difficulties of ensuring effective minority participation in a majority-based democratic system.

On the other hand, some governments make efforts to ensure the participation of indigenous peoples in development. For instance, Canada adopted a number of initiatives in this direction such as participation of indigenous peoples in environmental assessment and regulatory boards and in land claim settlement agreements. It further developed a regional partnering approach to increase the opportunities for indigenous peoples' employment. New Zealand has launched a capacity building program designed to assist Whanau, Hapu, and Iwi Maori communities to identify needs and develop initiatives to achieve long-term economic development.[42]

- In the eastern Amazon area of Brazil, the influx of settlers and loggers, attracted by multi-million dollar development projects, are threatening the survival of several indigenous peoples, among them the Awa. Elsewhere in the Amazon area, several indigenous peoples are asking for the demarcation of their traditional homelands according to existing legislation, and the return to lands from which they had been forcibly removed in previous years, among them the Kayabi of the Baixo Rio Teles Pires area and a group of Krahö Indians in Maranhäo state.[43]

[41] Report of the 18th session of the Working Group on Indigenous Populations (E/CN.4/Sub.2/2000/24), para. 31, and Report of the 19th. Session of the WGIP (E/CN.4/Sub.2/2001/17), para. 87 and 106.

[42] Report of the 19th. Session of the WGIP (E/CN.4/Sub.2/2001/17), para. 39.

[43] See the website of Conselho Indigenista Missionario: www.cimi.org.br.

- In British Columbia the Secwepemc first nation are struggling to protect their traditional land, which they use for multiple subsistence activities, from the planned expansion of a ski resort, on the basis of their Aboriginal Title recognized by the Supreme Court of Canada in 1997.[44]
- In Botswana and Namibia the Bushmen, numbering around 80,000 have been the victims of discriminatory practices and their survival as a distinct people is endangered by official assimilationist policies. Of particular concern is the fact that many groups have been dispossessed of their traditional lands to make way for game reserves and national parks. One non-governmental organization states that "unless fundamental rights such as that to land ownership are recognized urgently, the situation of the Bushmen will deteriorate further."[45]
- The hunter-gatherer forest dwelling Ogiek of Kenya who number approximately 30 000 people countrywide have long suffered dispossession of their land. They have challenged in the courts the government's continued intention to excise large parts of land for private development from areas traditionally occupied and held by the Ogiek.
- Malaysia has three main groups of indigenous peoples: the indigenous peoples of Sabah and Sarawak, who are covered by the 'special provisions' in the Constitution and the Orang Asli who are not covered by these provisions, but by specific legislation and administration. Although the 1954 Aboriginal Peoples Act, amended in 1967 and 1974, establishes certain rights for the Orang Asli, it mainly contains provisions restricting their rights to control their own lives. Orang Asli can live in specific lands and reserves, but cannot own them; authorities have the right to order an indigenous community to leave an area or stay away; they can revoke or vary any declaration of an aboriginal reserve with no compensation; they have no obligation for compensation for the loss of land, nor any obligation for relocation of the indigenous community or allocation of alternative land. Thus, Orang Asli have no security over their lands, but are reduced to be tenants at will. State authorities do not grant or sell land to the Orang Asli and there is little help for them to obtain individual title to the lands in which they live. Even when their lands are gazetted (officially announced), they have few rights and security. Development schemes and use of land for plantations and logging purposes often prevail over Orang Asli occupation of a certain land. Several land scale planned settlements schemes proclaimed in the 1960 Land (Group Settlement Areas) Act are implemented in Orang Asli areas. Their rights over their lands are lost in development schemes, highway and dam projects. Also, Orang Asli do not have any exclusive rights to the natural

---

[44] Information presented to the author directly during fact-finding mission in Canada, March 2003.

[45] Statement by Survival International to the World Conference Against Racism, August 2001.

products growing in their areas. Moreover, matters concerning Orang Asli and their lands are decided without their participation.[46]

- Despite the existence of national legislation concerning the rights of indigenous and tribal peoples, the implementation of these laws in Southeast Asia has been far from satisfactory. Indigenous representatives from the Philippines complain about the slow pace of implementation of the Indigenous Peoples Rights Act, adopted in 1997 arguing that the government has not yet allocated funds for its operationalization. Also, it is argued that the Act fails to fully protect indigenous lands from mining and logging. According to section 56, 'property rights within the ancestral domain already existing and/or vested upon effectively of this act, shall be recognized and respected'. Thus, leases for logging and mining shall continue to exist, even if an area is identified as an ancestral domain under the Act.[47]
- Other legislation allows turning indigenous lands into national parks and reserves for the sake of eco-tourism. Once free to roam the forest and harvest some products to sell in the lowlands, the new legislation curtails this former freedom of indigenous people severely. Moreover, private interests and foreign investors have established their activities in indigenous regions. In 2001, indigenous representatives reported to the Working Group on Indigenous Populations that two of the country's biggest companies have refused to compensate victims of disasters caused by their mining activities. As indigenous representatives state it, despite many positive points of the Indigenous Peoples Rights Act, other legislation stops its effective implementation.[48]
- Serious issues regarding the non-recognition of, and failure to respect, the rights of indigenous and tribal peoples have been reported in Suriname. Indigenous and tribal peoples (Maroons), who together comprise around 75,000 persons or about 14 % of the total population, occupy the forested areas of the 'interior' and suffer various types of discrimination in the national society. The government's report to the World Summit on Social Development recognizes these peoples as stakeholders in natural resources exploitation in their traditional lands but concedes that their participation in decision taking in those issues 'needs to be improved'. Legally, the land they occupy is owned by the state, which can issue land property grants to private owners. Indigenous and tribal lands,

[46] Orang Asli are called all the indigenous groups of peninsular Malaysia. For more information, see Minority Rights Group, Forests and Indigenous Peoples of Asia, London: MRG, 1999.

[47] Dario Novellino, 'The Ominous Switch: From Indigenous Forest Management to Conservation- The case of the Batak on Palawan island, the Philippines', Marcus Colchester and Christian Erni (eds), Indigenous peoples and Protected Areas in South and Southeast Asia, IWGIA Document No. 97, Copenhagen 1999, 250–297.

[48] Human Rights Watch, Asia, The Philippines, Human Rights and Forest Management in the 1990 s, April 1996; also, Survival of Tribal Peoples, Mountains of gold, the mining threat to tribal peoples in the Philippines, Survival 1996; also see Victoria Tauli-Corpuz, 'The resistance of the indigenous peoples in Asia against racism and Racial Discrimination', (2001) 1 Indigenous Affairs, 43–53.

territories and resources are not recognized in law. Various indigenous and Maroon communities have been affected by mining (gold and bauxite) and logging activities carried out by national and foreign companies, without their prior consent or participation. As a result, numerous villages have had to relocate against their will and their environment has been disturbed, disrupting their traditional subsistence economy, their health, their social organization and their culture. Despite petitions to the national government and the Inter-American system of protection of human rights (Commission and Court), the indigenous and Maroon communities have not received the protection they require.[49]

The International Labor Organization handles representations made by indigenous peoples concerning alleged violations of Conventions 107 and 169. At a recent session, the ILO's tripartite committee found that, in view of the importance of collective ownership of land for certain indigenous and tribal peoples, decisions involving legislative or administrative measures that may affect the land ownership must be taken in consultation with them. When communally owned indigenous lands are divided and assigned to individuals or third parties, this often weakens the exercise of their rights by the community or people, and they may end up losing most, if not all, of the land.[50] Consultation has also been dealt with within the context of a number of situations involving displacement for the purposes of development projects, particularly in a number of Latin American countries as well as in Asia. In each instance, one of the primary concerns of both the tripartite committee and the Committee of Experts of the ILO has been the apparent lack or inadequacy of consultations with the indigenous peoples affected by these projects, and lack of protection of displaced persons. The Committee expressed concern that the burden of such projects should not fall disproportionately on the tribal people inhabiting regions where these projects take place. Measures should be taken to ensure that they are provided with adequate protection.

Indigenous farmers and hunter-gatherers in forest environments are caught up in this maelstrom of change, and they often become uprooted and displaced, virtual 'development refugees', increasing the ranks of migrant laborers both within as well as across national boundaries. Millions of indigenous peasants have thus become itinerant agricultural laborers and migrants to large urban centers, sometimes also across international boundaries.

---

[49] Forest Peoples Programme, Failure of the Republic of Suriname to Recognize, Guarantee and Respect the Rights of Indigenous and Tribal Peoples to Lands, Territories and Resources, to Cultural Integrity and to be Free from Racial Discrimination. Formal Communication made Pursuant to Commission on Human Rights Resolution 2001/57.

[50] Report adopted by the Governing Body at its 271st Session (March 1999), regarding the representation made by the General Confederation of Workers of Peru, alleging non-observance of Convention No. 169 by Peru. See also Report adopted by the Governing Body at its 271st Session (March 1999), regarding the representation made by the Bolivian Central of Workers, alleging non-observance of Convention No.169 by Bolivia.

Indigenous communities and human rights organizations are working together to protect the lands to which they have a claim according to international and national legal standards. A landmark case in this direction is the decision of the Inter-American Court of Human Rights against the State and in favor of the Awas Tingni indigenous community in Nicaragua, concerning an "effective mechanism for official delimitation, demarcation and titling of the indigenous communities' properties, in accordance with the customary law, values, usage and customs of these communities".[51]

Similar judgments have been made by the courts in other states as well, so that indigenous land rights can, and indeed are, in some cases protected by favorable legal and court action. Still, these are exceptional cases, because generally indigenous communities do not have easy access to the judicial system and in a number of countries these remedies are not available to the indigenous at all. It therefore appears that in the future efforts must be made to improve access to the judicial system by indigenous communities and to reform the legal systems when indigenous peoples are denied access to legal recourse.

But even when laws are in principle available to the indigenous, these are not always implemented in their benefit. Numerous states report on recent legislative activity by which indigenous rights are seemingly protected, but indigenous organizations also report that their implementation leaves much to be desired. How to implement existing legislation effectively is as important for the rights of indigenous peoples as the adoption of such legislation itself. Moreover, not all legislation governing the ownership, use and access to land and other natural resources is favorable to the protection of indigenous rights. In some countries recent legislation undermines traditional communal or tribal holdings and opens the way to their dispossession by third parties or other private or corporate interests.

5.5 Homelands and Territories

While access to land for productive purposes (agriculture, forestry, herding, foraging) by individual members of indigenous communities is certainly of the greatest importance for indigenous people, there are other factors involved as well. Indigenous communities maintain historical and spiritual links with their homelands, geographical territories in which society and culture thrive and which therefore constitute the social space in which a culture can reproduce itself from generation to generation. Too often this necessary spiritual link between

[51] Inter-American Court of Human Rights, August 2001 (Judgment Summary and Order of the Inter-American Court of Human Rights. Issued 31 August 2001. In the case of The Mayagna (Sumo) Indigenous Community of Awas Tingni vs. the Republic of Nicaragua).

indigenous communities and their homelands is misunderstood by non-indigenous persons and is frequently ignored in existing land-related legislation.

Some scholars argue that the recognition of indigenous territorial rights is necessary for the full protection of the human rights and fundamental freedoms of indigenous peoples whereas others seem to fear that such recognition might undermine the unity and integrity of existing states. Nevertheless, in a number of countries such rights have indeed been legislated and experience suggest that national unity is not threatened by these developments.

After a decades-long struggle for legal redress concerning ancient land rights and aboriginal title, the Inuit people of northern Canada, who had linked land claims to territorial autonomy, negotiated a political agreement with the federal government, whereby they achieved the creation, in 1999, of the self-governing territory of Nunavut. Rather than weaken national unity, this arrangement has strengthened the federal structure of Canada and met the claims and aspirations of the Inuit people.[52]

In Panama seven indigenous peoples, the Ngöbe, Kuna, Embera, Wounaan, Bugle, Naso and Bri Bri, who together represent 8.3 % of the national population, are mostly concentrated in five legally constituted territorial units (*comarcas*) which make up almost 20 % of the country's total land area. These comarcas are semi-autonomous regions governed by local councils and traditional governors (*caciques*).[53]

How can and should existing states coexist with the notion of indigenous territories? Are these notions incompatible? To what extent is the idea of legally recognized indigenous homelands a necessary ingredient for the full enjoyment of the range of human rights by indigenous peoples? These are still open and debated questions, and answers will vary by region and country. While there are a number of practical experiences that illustrate the problems involved, more research is needed to address the particular issues, which are frequently controversial in public discourse. How can constructive arrangements be found between the legitimate concerns of states regarding territorial integrity and national unity, and the equally legitimate concerns of indigenous peoples regarding their collective survival qua peoples linked to the earth in myriad ways within an international system made up of sovereign states?

---

[52] See at: <www.nunavut.com>.

[53] Congresos y organizaciones indigenas de Panama, Informe de la Situacion de los Derechos Humanos de los Pueblos Indigenas de Panama, Panama, June 6, 2001 (Report presented to the Inter-American Commission of Human Rights).

5.6 Civil Wars and Violence

In some countries, indigenous peoples have been the victims of civil conflicts, involving guerrilla warfare, paramilitary units, military repression and other forms direct and indirect violence which has led to assassinations, forced disappearances, compulsory relocation, refugee flows, detention without due process, destruction of villages and entire communities etc. The human rights situation of indigenous people in the framework of civil conflicts past or present has been extensively documented, but the actual protection of their human rights involves complex and difficult issues. The Maya and Miskito of Central America, the Hmong in southeast Asia, the East Timorese, the Embera and Huaorani in South America, the Twa in east Africa have all, at one time or another, been hapless victims of civil or international violence and conflict. In some countries, 'Truth Commissions' were set up to elucidate the facts, in others special efforts at post-conflict reconstruction and reconciliation are being undertaken.[54]

- Numerous reports document the situation of the Embera in Colombia who are victims of the civil war between the government of Colombia and the FARC (Fuerzas Armadas Rebeldes de Colombia). Their leaders and spokespersons have been threatened, persecuted, arrested and killed by paramilitary groups, guerrillas, or members of the police or armed forces (not to mention drug-traffic related violence).
- In northern Ecuador the Shuar have suffered fall-out from the violence in neighboring Colombia and the 'war on drugs'. A fact-finding mission organized by a group of non-governmental organizations in July 2001 reports increasing militarization in the area, environmental destruction, kidnappings, disappearances and killings of individuals and a general deterioration of social, economic and cultural conditions in the indigenous communities.[55]
- After 30 years of civil war in Guatemala a peace accord was signed in 1996 in which the rights and culture of the indigenous Maya people were agreed upon. In September 2001 the United Nations Mission to Guatemala (MINUGUA) published a Verification Report, which concludes as follows:

"After more than seven years since the Global Agreement on Human Rights took effect and almost five since the signing of the Agreement on a Firm and Lasting Peace, the Mission feels that little progress has been made in one of the most important areas for the consolidation of a democratic State that is inclusive,

[54] In Guatemala the peace accords of 1996 established a truth commission which documented massive atrocities committed by the military against the indigenous Maya during 30 years of civil war, while the government of Chile set up a "Commission on Historical Truth" in 2001, chaired by a former president of the republic, to study the situation of the country's indigenous people. (It is to present its report in October 2003).

[55] Communication "Taruka Report", presented to the author by a delegation of Ecuadorean Indians.

participatory and non-discriminatory. The ethnic diversity of the Nation is not yet recognized and valued as one of the greatest riches and the indigenous peoples continue to be subject to strong racial, ethnic and cultural discrimination which deprives them of the enjoyment of their basic human rights.

"The excluding economic and social models, as well as the centralization of public investment have led to a situation in which broad sectors of the population living in rural areas are subjected to conditions of poverty and extreme poverty, especially the great majority of the country's indigenous peoples. At the same time, the high concentration of indigenous population in regions where the armed confrontation was the most intense has made these peoples the victims of the worst consequences of the conflict which devastated Guatemala for more than 30 years."[56]

- In the wake of the genocide in Rwanda in the early 1990s, members of the Batwa tribe, considered as the indigenous people of the country, have suffered from persecution and reprisals. Some are languishing in jail, accused of acts of genocide in Rwanda. The historic links between Batutsi and Batwa made them vulnerable to attack by the Bahutu during the period of genocide. Between December 1993 and March 1994 at least 11 Batwa settlements were burned to the ground, the people attacked and some killed.[57]
- In the Central Highlands between Viet Nam, Laos and Cambodia the indigenous Montagnard/Degar people have been involved in conflicts and become the victims of human rights abuses over a period of many years. Montagnard asylum seekers from Viet Nam have reportedly been expelled from Cambodia and may be at risk. The OHCHR/Cambodia and the UNHCR have monitored the situation closely and the latter is involved in negotiations with Cambodia and Viet Nam concerning Montagnard refugees. Around 600,000 highland refugees are reportedly scattered inside Laos and Thailand without any UN protection, facing arrests, detention and brutality. The 2001 U.N. Report of the situation in human rights in Cambodia states that persons from the Central Highlands were driven out from Viet Nam for fear of persecution, because most of them 'had taken part in Viet Nam protesting the continuous confiscation of land by the Vietnamese Government since the end of the war in 1975'.[58]

In a number of Southeast Asian countries the resettlement of indigenous peoples and ethnic minorities in highland areas is the result of 'sedentarization'

[56] United Nations Verification Mision in Guatemala. Verification Report. The Indigenous Peoples of Guatemala: Overcoming Discrimination in the Framework of the Peace Agreements, September 2001.

[57] Special Rapporteur on the Human Rights and Fundamental Freedoms of Indigenous People, First Annual Report to the Commission on Human Rights, E/CN.4/2002/97.

[58] See http://www.hrw.org/press/2000/05/cambodia-0520.htm; Statement by Dr. Vuag Pobzeb of the Lao Human Rights Council, presented in the 2001 UN working group on indigenous populations. Cf. Situation of Human Rights in Cambodia, Report of the Secretary-General, A/56/209, (26 July 2001), para. 73.

policies designed to further certain rural development objectives, without sufficient regard to the needs and aspirations of the communities thus affected. Regarding Viet Nam, CERD has expressed concern 'about the alleged population transfer to territories inhabited by indigenous groups, disadvantaging them in the exercise of social, economic and cultural rights.'[59] Despite the government of Viet Nam's repeated denial that there is any racial discrimination in this country, there continue to appear reports of the use of force, land confiscations, forced sterilizations, killings and other forms of persecution of indigenous people in the area.[60] Similarly, the identity of the indigenous peoples, their cultures and their traditional way of living in Cambodia are reportedly seriously at risk, due to violations of land and citizenship rights. Some of the human rights abuses in the region occur within the frame of internal armed conflict, where anti-terrorism legislation and the granting of emergency powers not only delay the peaceful negotiation of differences but tend rather to further human rights abuses.

In several North African countries (mainly Algeria and Morocco), the Amazigh (also referred to as Kabyles, Touareg, Berbers), who consider themselves as indigenous to these countries, have been asking for the official recognition of their language, culture and identity, as well as the full enjoyment of their civil and political rights. After a rebellion in 1990, which actually started in Niger, the Touareg of Mali entered into a peace treaty with the government in 1991, followed by another one in 1992, to allow them regional self-governance and internal democracy, enabling the government to grant autonomy to the northern areas of the country occupied by the Touareg.[61]

While in Africa there is no consensus on the use of the concept 'indigenous', the African Commission on Human Rights has argued for the protection of the human rights of specific sectors of the population, as for instance, in a decision that involves the protection of rights of black populations in Mauritania. The Commission found that "language is an integral part of the structure of culture; it in fact constitutes its pillar and means of expression par excellence. Its usage enriches the individual and enables him to take an active part in the community and in its activities. To deprive a man of such participation amounts to depriving him of his identity."[62]

In dealing with human rights issues raised by indigenous and human rights organizations, it should be noted that while it is individuals who suffer abuses, these violations generally occur because they are members of distinct indigenous communities, tribes or peoples, and that indigenous collectivities are often singled

---

[59] CERD, Concluding Observations of the Committee on the Elimination of Racial Discrimination: Viet Nam. 15/08/2001.CERD/C/59/Misc.21/Rev.3, para. 14.

[60] CERD/C/SR. 1481, para. 4. See reports of the Montagnard Foundation presented to the UN; also see International Commission of Jurists, Report by the Western Australian Branch of the International Commission of Jurists concerning sterilisations of Ethnic Montagnards in the Central Highlands of Vietnam: October 2000.

[61] E/CN.4/2002/97.

[62] *Ibid.*

out as victims of abuses such as land loss, environmental destruction, forced displacement, imposed assimilation, ethnocide and so forth. The collective rights of indigenous peoples are thus often placed at the center of the issues that come to the attention of international agencies such as the WGIP, the ILO, the CERD and the UN Human Rights Commission. These issues are being taken up increasingly by regional bodies as well, for instance the Inter—American Commission on Human Rights and the African Human Rights Commission. Thus emerges a pattern of collective discrimination of entire communities, tribes and peoples.

Indeed, numerous formally recognized legal rights of indigenous peoples are often not fully implemented in practice, either in the courts by way of final adjudication determined by the judiciary, or as a result of new legislative acts which in fact weaken or reduce previously legislated rights. Collective discrimination may occur in democratic states where the rule of law prevails. This concern has been expressed by indigenous participants at the WGIP.[63] In relation to this process in the case of Australia, the Committee on the Elimination of Racial Discrimination (CERD), recommended that "close scrutiny continue to be given to any other proposed state and territory legislation to ensure that protection of the rights of indigenous peoples will not be further reduced."[64]

The vulnerability of land and other rights is also a problem besetting the 'Small Peoples' in the Arctic regions of Asia, an issue that has been taken up by the *Committee on the Elimination of Racial Discrimination* (CERD). In these regions, indigenous children face discrimination and other severe problems. In 1999, the Committee on the rights of the child expressed its concern for the living conditions of indigenous peoples in the north and their access to health, educational and other social services. The Committee referred to the growing incidence of societal discrimination against children belonging to ethnic minorities, including indigenous peoples and asked the Russian federation to take all appropriate measures to improve the situation. Notwithstanding the extensive rights of Samis in the Scandinavian countries, Sami women and children still face discrimination, as indicated by the concern expressed by the *Committee for the Elimination of Discrimination against Women* (CEDAW).[65]

[63] Report of the 18th session of the WGIP, E/CN.4/Sub.2/2000/24, Para. 165.

[64] Concluding Observations by the CERD: Australia. 19 April 2000. CERD/C/304/Add.000. (Concluding Observations Comments).

[65] CERD/C/SR.1246 of 5 March 1998; CRC/C/15/Add.110, of 10 November 1999, para. 65; A/56/38,paras.319–360 of 31 July 2001, para. 356. Also Russian NGO intervention at the meeting of the WGIP in July 2001.

5.7 Profiles

5.7.1 *Philippines*

About 140 indigenous ethno-linguistic groups, representing 15–20 % of the total population of 80 million, are present in more than 50 of the country's 78 provinces. They have continuously lived as communities in communally bounded and defined territories, which they have occupied from time immemorial. The legal framework in which indigenous rights must be considered under the Constitution is the Indigenous Peoples Rights Act of 1997, which also established the National Commission on Indigenous Peoples.

For poor indigenous farming communities crucial land rights are addressed by filing legal claims to their own ancestral domains and titles. The process is cumbersome and indigenous representatives perceive that the business interests of private enterprises, which over the years have encroached upon their ancestral domains, are more protected than their own rights based on land use and continuous occupation. High poverty rates and the lack of basic social services force many indigenous to migrate to poor urban areas where the situation of women and children is of particular concern.

Large scale economic activities such as logging, open-pit mining, multi-purpose dams, agribusiness plantations, and other development projects, are having long-term devastating effects on the livelihood of indigenous peoples and their environment. These activities are often carried out without their prior, free, informed consent as the law stipulates. Communities resist development projects that destroy their traditional economy, community structures and cultural values, a process described as 'development aggression'. Indigenous resistance and protest is frequently countered by military force, involving numerous human rights abuses, such as arbitrary detention, persecution, killings of community representatives, coercion, torture, demolition of houses, destruction of property, rape, and forced recruitment by the armed forces, the police, or so-called paramilitaries.

5.7.2 *Chile*

The Mapuche in southern Chile, who make up more than half of the country's indigenous population, have lost most of their ancestral territory as a result of the expansion of large agricultural enterprises and the privatization of their land undertaken aggressively by the military dictatorship (1973–1991) and continued to this day. The government's program to buy land for the Mapuche, under Law 19.253 (adopted in 1993) has been slow and insufficient. Currently small individual Mapuche farms are scattered islets in a sea of large commercial estates. Militant Mapuche organizations insist that more attention be given to their needs and they have staged protests to draw attention to their longstanding and neglected claims. Government authorities have expressed their willingness to address legitimate demands, but they have also cracked down on Mapuche protesters, accusing them unjustly of engaging in violent terrorist acts. Numerous organizers are in jail and the local tensions have created a climate of political confrontation at the national level which does not help the cause of indigenous rights at the present time when a proposal to reform the constitution in order to recognize such rights is stalled in the national congress. The demand for social services for the Mapuche and other indigenous peoples has not been met, and the HDI of the Mapuche (particularly women) is systematically lower than the national average. While intercultural and bilingual education is part of government policy, its results to date have been minimal and indigenous organizations complain that their educational services are below par.

In arid northern Chile small indigenous Aymara and Atacameno communities are being denied the necessary water resources for survival, which large mining interests are able to appropriate according to the law. Their poverty index is high and many younger people move to the cities in search of jobs and income. The preservation of their cultural and linguistic identity is severely threatened.

Easter Island in the Pacific, inhabited by the Polynesian people of Rapa Nui, was incorporated into Chile in the nineteenth century. The remaining indigenous population of about 2,000 people is being swamped by the increasing immigration of outsiders who have taken over most of their land. They are asking for more local autonomy and a direct relationship with the national government, rather than with a provincial one as is now the case.

5.7.3 Mexico

The Zapatista uprising in 1994 put the issue of indigenous rights squarely on the national agenda, but a peace accord, signed in 1996, remained on paper. In 2001 the new government passed a constitutional reform on indigenous issues that deviated from the agreements and further stalled the peace process. Subsequently, in 2003 a number of indigenous municipalities, which earlier had declared their autonomy, created parallel government structures to promote their own vision of development as set out in the peace agreements, whereas the government decided to set up a new National Commission for the Development of Indigenous Peoples and a National Institute of Indigenous Languages. At the local level conflicts over land and resources often turn into acts of violence, and indigenous persons frequently become the victims of a corrupt and biased judiciary system. Indicators of social wellbeing are much lower in the indigenous rural communities than in non-indigenous urban areas, leading to massive migrations of Indians to other parts of the country and across the border to the United States. If carried out as announced, the Puebla Panama Plan of the governments of Mexico and Central America may further affect the potential of indigenous communities to survive as distinct cultural entities in a globalized world. Indigenous organizations demand not only respect for their culture and languages, but also of their rights to self-determination and autonomy as well as full participation in the political and social process.

5.7.4 Guatemala

More than half of the national population consists of indigenous, mainly Maya, people who are now officially recognized in the Peace Agreement on the Identity and Rights of Indigenous Peoples, signed in 1995 after 30 years of brutal civil war. Access to land and resources is nevertheless still the main problem faced by indigenous communities, who also continue to be the victims of discrimination and marginalization. Indigenous identity, extreme poverty, poor access to educational and health services are all closely related. One of the areas in which discrimination against indigenous people is especially strong is the administration of justice system, which despite a major effort made by the government in recent years, is still cumbersome and inefficient. Social conflicts are often criminalized, creating dissatisfaction with the judiciary among the indigenous communities. Lynchings of suspected offenders have become commonplace in local communities where the reach of the law is absent. Local police forces are still in many places controlled by members of the paramilitary groups that committed brutal atrocities during the war, and despite the peace agreements and a supervisory mechanism set up by the United Nations, human rights violations are again on the increase.

Numerous social and human rights organizations, including a vibrant Maya movement, have been working actively over the years on an agenda aiming at the full respect and participation of indigenous peoples within the national society. Several indigenous ministers in the government are promoting public policies in the field of indigenous education and culture; Maya languages are spoken by half of the population, and bilingual intercultural education is an objective of the formal educational system.

5.7.5 Japan

The Ainu, the original indigenous inhabitants of the island of Hokkaido, were formally incorporated into the Japanese state in the nineteenth century. Official government policy was to integrate them into Japanese society and culture, a process that over the decades led to the almost complete loss of Ainu ethnic identity. The Ainu soon became a minority on their own ancestral territory.

The first reaction to preserve their vanishing culture was undertaken about 20 or 30 years ago by Ainu activists concerned about the loss of their language, traditions and identity, which formed the Ainu Association of Hokkaido. They finally achieved a long-hoped for result: the passage of the Ainu Cultural Promotion Law of 1997. While recognizing its merits, Ainu activists state that the Law does not satisfy their aspirations, because it does not formally recognize their social and cultural rights as an indigenous people.

A landmark case for the Ainu was achieved in the Sapporo District Court, which decided in 1997 that the building of the Nibutani Dam on the Saru River illegally affected traditional sacred and burial sites of Ainu communities. Ainu people, particularly women, report incidents of discrimination against them in daily activities. Nowadays Ainu cultural activists undertake a series of activities at the community level to preserve knowledge of the Ainu language and their arts and traditions. They hope for more government support and understanding by the rest of society (as well as other Ainus), because they are aware that their survival as an indigenous people is at stake in this process.

5.7.6 Canada

Though they represent only a small fraction of the total population, Canada's First Nations claim that as the original inhabitants of the country they have been marginalized and discriminated against in the Canadian Federation. Having lost their former sovereignty, they have had to endure endless pressure on their territories, land and resources, a condition which has transformed most of them into wards of the state, depending on governmental subsidies and social services for their survival. Many First Nations have gone to the courts to defend their aboriginal treaty rights and their aboriginal land title claims, where sometimes they have indeed received favorable judicial decisions. They demand equal treatment with the other two 'founding nations' of Canada and claim the right to self-determination. Some of them are involved in struggles to retain control over and access to some of the last natural resource domains on the continent (forests, lakes, streams, fisheries) that are being coveted by powerful economic interests. The insistence of the First Nations on their treaty rights is essential for their survival as culturally distinct peoples.[66]

5.8 Education and Culture

Indigenous peoples tend to maintain a cultural distinctiveness that distinguishes them clearly from other groups in society and from those sectors that are usually identified by the concept of 'national culture'. There are numerous features associated with this cultural distinctiveness, the main one being the use of their own language, which is not only a medium of communication, but also a crucial

---

[66] The brief profiles of Philippines, Chile, Guatemala, Mexico, Japan and Canada are based on the authors observations during fact-finding missions to these countries in 2002–2003.

element in the structuring of thought processes and in providing meaning to the natural and social environment of any person. A language community is also an epistemic community, that is, it links people through their participation in a common medium and in shared understandings. Indigenous language communities provide their members with the full range of cultural meanings attached to the use of a shared idiom. Most indigenous languages are very ancient and while they have undergone changes—just as any other language—they are transmitted from generation to generation and thereby help preserve the continuity of a language community and its culture.

Language rights are an essential element of the cultural rights that all persons enjoy under international human rights standards. The right to one's own language pertains not only to individuals but also to communities, nations and peoples. If a language community as such is denied the collective and public use of its language (for example, in schools, the media, the courts, the administration) then any individual's right to this language is severely curtailed. Therefore, language rights are nowadays proclaimed as human rights, which entail respect, protection and promotion by others and especially by public authorities. Numerous states have now adopted legislation concerning the protection of regional, minority or indigenous languages. For example, in New Zealand, the Education Act ensures funding for Maori pre-schools, primary schools, secondary schools and universities. The impetus for this came from Maori mothers insisting that Maori reclaim the education of their children from birth through to adulthood.

In historical perspective, however, state policies have not always recognized or protected the languages spoken by indigenous peoples or linguistic minorities. On the contrary, the intention of official linguistic, educational and cultural policies has often been the assimilation of such groups into the national mainstream, thus leading to language and cultural loss. It has only been in recent years that these processes have been seen as being in violation of the human rights of the members of such linguistic communities, and they have sometimes been considered as a form of ethnocide.[67]

In some countries indigenous languages are recognized as national languages, at least in the regions in which they are widely used, and sometimes they have been accorded official status of some kind or another. In other cases, they may no longer be actually repressed but only tolerated as a private medium of communication but are not accorded any official status. In numerous indigenous linguistic communities around the world, it is common to find members of the older generation who maintain their language whereas youth and children are more prone to suffer language loss, particularly when assimilationist policies are carried out. Article 30 of the Convention on the Rights of the Child is clear: "In those States in which ethnic, religious or linguistic minorities or persons of indigenous origin exist, a child belonging to such a minority or who is indigenous shall not be denied

[67] Ethnocide is a process of cultural change and destruction as a result of specific policies that undermine a cultural community's ability for self-preservation.

the right, in community with other members of his or her group, to enjoy his or her own culture, to profess and practice his or her own religion, or to use his or her own language."

The denial of the right to practice one's own culture, religion or language may take many forms. Often when the social and institutional environment is unfavorable for the preservation and development of indigenous cultures and languages, this right is in fact denied even when there is no formal prohibition or restriction involved.

The use of the mother tongue in education and public communications is an important issue for indigenous communities. In contrast to the formerly widely extended and dominant idea of formal schooling as an instrument of assimilation and acculturation, through which indigenous children learn to speak the official national idiom and replace their native tongue, current thinking on the subject tends more towards the opposite direction. Bilingual and intercultural schooling has become the object of educational policy for indigenous communities in many parts of the world. Specialists in education agree that early schooling in both the native mother tongue and the official language of the state is of great benefit to indigenous children, who may become proficient in the vehicular language of the wider society without losing their vernacular idiom.

Nevertheless, despite the best of intentions, the teaching of native tongues in schools has its difficulties. In the first place, many indigenous languages lack their own alphabets and do not have a written tradition. Secondly, the formal teaching of the vernacular tongue and of the vehicular or official idiom as a second language requires special training and pedagogical skills which indigenous teachers often lack. In Mexico, for example, as well as in other Latin American countries where official bilingual education in indigenous areas has a history of some decades, the output level of students in bilingual schools is still below that of the national average. Furthermore the preparation of textbooks and teaching materials in indigenous languages usually lags behind those in areas where the national or official language is taught exclusively. In many countries school administrations (either public or private) are not set up to handle indigenous bilingual education effectively. To that extent, the indigenous right to education in their own languages is not being adequately implemented and requires serious attention in the future.

Even more problematic is the idea of multicultural or intercultural education, because this involves not only local schools but also the regional and national school systems and the educational philosophy of any country where there are indigenous peoples. The notion of multicultural and intercultural education leads to a complete revision of educational contents and methods in countries where it is applied. It basically means that the cultural diversity of the country be reflected in the curriculum and the preservation and promotion of cultural diversity become an objective compatible with democratic governance and the enjoyment of human rights by all. In some cases this approach will require the revision of traditional ideas held by majority or dominant cultural groups about national culture and identity. Indigenous peoples' organizations often need to remind the world that their own cultural specificities are also contributions to a universal culture and not mere relics of a disappearing past. The rights of indigenous peoples to culture and education (the whole gamut of cultural rights, in fact) include the right to the enjoyment and protection of their own cultures in a wider, multicultural world.

5.8.1 Multiculturalism

The preservation of indigenous cultures (including tangible and intangible elements, arts and artifacts, traditions, knowledge systems, intellectual property rights, ecosystem management, spirituality and so on) is an essential component of a comprehensive indigenous human rights package, but in fact the preservation of indigenous cultures is not a natural process at all. The contrary is more likely, because as has been well documented in the specialized literature on the topic, public policies have frequently been designed to eliminate and transform indigenous cultures because their existence has often been considered as detrimental to the idea of national integration and development. Many countries adopted specific policies to 'assimilate' indigenous peoples into the wider 'national' culture within the framework of cultural and social modernization. While such ideas no longer command the support they used to have, and whereas more and more States adopt positions favorable to multiculturalism, there are still numerous cases in which the cultures of indigenous peoples are under strong outside pressures to change, when they are not actually on the verge of extinction.

The idea of multiculturalism does not imply the artificial preservation of indigenous (or tribal) cultures in some sort of museum, but only the right of every human community to live by the standards and visions of its own culture. Certainly cultures change over time, but whether there will ever be one universal culture or any number of interrelated local, regional, ethnic, and national cultures, only time will tell. In human rights terms, it is clear that cultural rights pertain to every individual, yet these rights can only be fully enjoyed by all persons in community with other members of the group. Thus indigenous peoples require guarantees that their cultures will receive the respect and consideration that other groups in society also enjoy, and that they will have the freedom to develop their cultural creativity in communion with other members of their group. At the international level, these issues have been taken up by UNESCO and by WIPO with regard to the cultural heritage and intellectual property of indigenous peoples.

The cultural rights of indigenous peoples are also addressed in a number of national legislations, though not always with the clear intent to promote and enhance them. For example, in the Philippines, the Constitution includes several provisions concerning the rights of the 'cultural communities', and Article IV states that 'the State shall recognize, respect and protect the rights of indigenous cultural communities to preserve and develop their cultures, traditions, and institutions'. Most Latin American constitutions now have provisions affirming the right of indigenous communities to preserve their own cultures and stating the obligation of governments to protect them. But here, as elsewhere, the devil is in the details. Implementing legislation and practice usually lags well behind the principles set out in the constitutional texts.

Since the 1980s a number of Latin American states have reformed their political constitutions, many of them recognizing for the first time indigenous peoples as deserving of special attention by the state and as holders of specific rights relating to land, language, culture, social organization and other features. Some texts are more specific on indigenous rights whereas others refer only to the protection and preservation of culture. The right of peoples to self-determination and autonomy is dealt with in only a few of these documents. Implementing legislation has been slow in coming and indigenous organizations tend to point out that in many instances the constitutional texts are not actually being complied with by the authorities.

It was pointed out above that indigenous cultures are closely linked to the concept of land rights and the occupation and possession of territorial homelands. A question frequently asked of indigenous peoples is whether their cultural identities can survive in a deterritorialized environment, that is, in dispersed settlements and urban centers where indigenous migrants live interspersed with non-indigenous populations. The answer to this question depends on particular circumstances and is contingent on the specific definition of indigenous identity in each case. It may be argued that to the extent that cultural rights are universal, they are not subject to any kind of territorial restriction. The right of any individual or group of individuals to preserve, practice and develop their own culture is not dependent upon territoriality but rather related to self-identification. Indigenous identities have indeed been diluted in the process of urbanization, but under certain circumstances the urban environment is favorable to the emergence of new kinds of indigenous cultural identities. The majority of Mapuche Indians live in Chile's capital, Santiago, where they have organized strong militant political, social and cultural associations based on their ethnic identity. Some Mexican indigenous migrant groups have formed ethnic associations that actually straddle the US-Mexican border in urban environments, turning into a new type of 'transnational community'. In the Philippines about half the population of Baguio City, the major urban center of the Cordillera, consists of immigrants from outlying indigenous villages, just as Quetzaltenango in Guatemala has a strong Kiche (Maya) identity. In other metropolitan centers ethnic neighborhoods are able to maintain indigenous identities over several generations.

How the linguistic, educational and cultural rights of indigenous peoples are being protected or not under varying circumstances is an empirical question that needs more comparative research. UNESCO has recommended that states take special measures to ensure the protection and promotion of indigenous cultures. The African Commission on Human and Peoples' Rights has set out guidelines that require states to take specific measures aimed at the promotion of cultural identity and the "awareness and enjoyment of the cultural heritage of national ethnic groups and minorities and of indigenous sectors of the populations".[68]

A major issue facing indigenous peoples is the protection of their cultural heritage, including traditional knowledge, skills and techniques of all kinds. Frequently the topic is dealt with under the label of 'intellectual property' but this concept has commercial connotations that are unrelated to the way indigenous peoples conceive of their culture. One study on the subject concludes that the distinction between cultural and intellectual property is, from indigenous peoples' viewpoint, an artificial one and not very useful. Industrialized societies tend to distinguish between art and science, or between creative inspiration and logical analysis. Indigenous peoples regard all products of the human mind and heart as interrelated, and as flowing from the same source: the relationships between the people and their land, their kinship with the other living creatures that share the land, and with the spirit world. Since the ultimate source of knowledge and creativity is the land itself, all of the art and science of a specific people are manifestations of the same underlying relationships, and can be considered as manifestations of the people as a whole.

Cultural heritage "is everything that belongs to the distinct identity of a people and which is theirs to share, if they wish, with other peoples. It includes all of those things which international law regards as the creative production of human thought and craftsmanship, such as songs, stories, scientific knowledge and artworks. It also includes inheritances from the past and from nature, such as human remains, the natural features of the landscape, and naturally-occurring species of plants and animals with which a people has long been connected."[69]

In 2,000 the UN organized an international seminar on the protection of the cultural heritage of indigenous peoples which adopted a set of principles and guidelines, among others the following:

1. "The effective protection of the heritage of the indigenous peoples of the world benefits all humanity. Its diversity is essential to the adaptability, sustainability and creativity of the human species as a whole.
2. To be effective, the protection of indigenous peoples' heritage should be based broadly on the principle of self-determination, which includes the right of indigenous peoples to maintain and develop their own cultures and knowledge systems, and forms of social organization.
3. Indigenous peoples should be the source, the guardians and the interpreters of their heritage, whether created in the past, or developed by them in the future.

[68] African Commission, General Guidelines Regarding the Form and Contents of Reports to be Submitted by State Members regarding the Meaning, Scope and Weight of the 'Rights of Peoples' Recognised by Articles 17(2), 19 to 20 of the Charter. 1990: 417–8.

[69] Erica-Irene Daes, "Study on the protection of the cultural and intellectual property of indigenous peoples", E/CN.4/Sub.2/0993/28.

4. Recognizing, respecting and valuing their customs, rules and practices for the transmission of their heritage to future generations is essential to indigenous peoples, their identity and dignity.
5. Indigenous peoples' ownership and custody of their heritage should be collective, permanent and inalienable, or as prescribed by the customs, rules and practices of each people."[70]

- "A song, for example, is not a "commodity", a "good," or a form of "property," but one of the manifestations of an ancient and continuing relationship between the people and their territory... [and therefore] it is inconceivable that a song, or any other element of the people's collective identity, could be alienated permanently or completely."[71]
- The Cultural Heritage Act No. 3501 (1979) of Ecuador is applicable to all that indigenous peoples themselves regard as "recurrent and valid means of expression and identification of their culture".
- In Sweden the Sameting, an elected governmental authority of Sami people established in 1992, decides how to allocate national funding for the promotion of Sami culture and the Sami language. Sami cultural heritage is included in the curricula of all Swedish schools, and Sami communities are free to establish their own schools. All abandoned sacred and ceremonial sites are protected by Swedish law; however, Swedish law does not distinguish between the intellectual property rights of Sami and other citizens.

5.9 Convention on Biological Diversity

The wealth of knowledge that indigenous communities possess concerning the environment, the plants and animals of their traditional habitat and the multiple uses that such knowledge can be put to is one of the principal assets of indigenous cultures. During colonial times and up to fairly recently such knowledge was widely used and shared, particularly in connection with traditional herbal medicine, nutrition, colors used for weaving textiles and making handicrafts, hunting, fishing and gardening etc. As a result of the dissemination of the technological and scientific achievements of the post-industrial society, however, much of this knowledge was neglected and discarded, when not actually rejected and its use forbidden by authorities of all stripes who wanted to 'modernize' their countries. For example, herbal specialists were in some parts forbidden to practice their skills openly, or access to certain locations necessary for such practice was refused to members of indigenous communities on legalistic grounds.

[70] E/CN.4/Sub.2/2000/26.

[71] Ibid. See also E/CN.4/Sub.2/0995/26.

Indigenous people are demanding respect for their traditional knowledge and the freedom to use the products of their environment according to custom. Moreover, some indigenous communities have been able to merchandise such products and thereby increase their incomes. More recently, multinational corporations have discovered the commercial potential inherent in much of this traditional wealth, and the world race is on to patent, privatize and appropriate what has been part of the cultural heritage of indigenous peoples from time immemorial. Once again, the indigenous are being dispossessed of their legitimate collective property, only this time within the framework of multilateral commercial and financial agreements arranged by corporations, universities and governments. This dramatic reversal of fortunes is deeply hurting indigenous peoples the world over, and only through concerted international action will they be able to save what little is left of their cultural heritage.

One such favorable development is the Convention on Biological Diversity (1992), which has so far been ratified by 134 states. The Convention provides that states shall:

"… respect, preserve and maintain knowledge, innovations and practices of indigenous and local communities embodying traditional lifestyles relevant for the conservation and sustainable use of biological diversity and promote their wider application with the approval and involvement of the holders of such knowledge, innovations and practices and encourage the equitable sharing of the benefits arising from the utilization of such knowledge, innovations and practices" and "protect and encourage customary use of biological resources in accordance with traditional cultural practices that are compatible with conservation or sustainable use requirements".

Nevertheless, there is still a long way to go from the text of this Convention to its actual implementation by states at the national level.

5.10 Social Organization, Local Government, Customary Law

Cultural identities are sustained not only by a discrete list of 'elements' that every member of a cultural group 'carries along' as he/she goes through life. In fact, these elements may vary from individual to individual and they may, and frequently do, change over time. So it is not the contents of a culture which defines any group's identity. It is rather in the field of social organization that identities are wrought and sustained. To the extent that a system of social relations defines the identity of each individual member and his/her link to the group as a whole, the social institutions and relationships characteristic of a given community are the necessary frame of reference for any culture to thrive. Indigenous communities know this well, because when they claim the right to maintain their social organization in the face of the pressures of the wider society, they are actually appealing for the preservation of their culture.

Too often the larger society has taken the stance that indigenous social institutions are contrary to the national interest or, worse, are morally reprehensible. This position was taken for a long time by the dominant institutions in colonial empires. The question is frequently debated whether adherence to indigenous communal institutions may lead under certain circumstances to the violation of individual human rights (for example, the rights of women and girls).

Local community organization is often upheld by adherence to a generally accepted system of customs and mores or customary law, which in numerous countries is not accorded any formal legal recognition and may in fact be considered as competing with the formal state legal system. Do community members who accept the norms of unwritten customary law stand in violation of a country's legal system? Does the application of customary law violate nation-wide legal norms? Yet what about situations in which the application of positive law entails a violation of community norms and customs? Might that not constitute a violation of human rights as well?

These issues are dealt with in different ways by individual states (and by different scholars) and the various solutions run from some form of accepted legal pluralism to the absolute rejection by the official legal system of any kind of indigenous customary law, with a number of possibilities in between. Under what circumstances might the application of indigenous legal systems (customary law) threaten internationally accepted standards of individual human rights? And conversely, under what circumstances could the limitation or elimination of indigenous customary law violate the human rights of members of indigenous communities? These are complex issues about which there is much debate and little agreement, which need to be addressed objectively and without bias.

Since time immemorial, local communities have evolved some form of local government within the structure of a wider polity into which they have been integrated as a result of historical events. Indigenous communities are no exception. Throughout history, local communities have struggled to defend their autonomy against outside encroachment, sometimes successfully, sometimes not. To the extent that indigenous people were incorporated into state structures not of their own choosing during times of colonization or the expansion of the modern nation state, their local forms of government were modified or adapted to suit the interests and needs of the state, creating tensions that have often led to conflict and violence.

Indigenous organizations seek to preserve or regain the right to local (and sometimes regional) self-government; they consider this right as part of the fundamental freedoms which international law accords to all peoples. Through negotiations and treaties, constitutional reform or special legislation, indigenous peoples have been able in numerous instances to establish agreements with states regarding this right to self-government. In other cases, however, this has not been possible, and national—or regional-level government units still take it upon themselves to administer the affairs of indigenous communities. Indigenous affairs ministries, departments or bureaus often have specific mandates to that effect and local indigenous governments need to deal with these institutions rather than with those of the national political or administrative system in general. Indigenous organizations may consider this to be a form of discrimination, whereas governments argue that such arrangements are designed for the protection of indigenous people themselves, in keeping with their best interests (as defined by the state).

Recognizing these issues, the *Draft Declaration on the Rights of Indigenous Peoples* states in article 33: “Indigenous peoples have the right to promote, develop and maintain their institutional structures and their distinctive juridical customs, traditions, procedures and practices, in accordance with internationally recognized human rights standards”.

5.11 Poverty, Levels of Living, Sustainable Development

As already noted, indigenous people are very often found among the poorest strata in society, their levels of living are considered to be sub-standard in many respects. Studies have shown high levels of infant mortality, lower than average nutritional levels, lack of public services, difficulty of access to social welfare institutions, lower than average delivery of the services provided by such institutions, inadequate housing and shelter and other indicators associated in general with the idea of human development.

Many states have recognized these problems and promote special policies and measures designed to improve the levels of living of indigenous people. In other areas public policies are not oriented in this direction and the needs of indigenous populations have been neglected. Numerous statements made by indigenous representatives to the WGIP over the years, and other information gathered by independent research bodies, confirm this tendency. For instance, the Committee on Indigenous Health of the Indigenous Peoples’ Caucus expressed its concern at the nineteenth session of the WGIP that the gap between the health of indigenous peoples and the rest of society is widening, despite all efforts by national governments and international agencies.[72]

The development of Latin America has been highly unequal, and the benefits of economic growth are concentrated at the upper end of the social and economic scale. While poverty and extreme poverty are widespread all over rural and urban Latin America, the indigenous peoples are mainly concentrated at the lower levels. The World Bank reported in the 1990s that the living conditions of the indigenous people were abysmal, and that their poverty was persistent and severe, especially when compared to those of the non-indigenous population.”[73]

---

[72] Statement of the Committee on Indigenous Health (COIH) to the WGIP, July 2001.

[73] George Psacharopoulos and Harry Anthony Patrinos (Eds.). 1994: *Indigenous People and Poverty in Latin America. An Empirical Analysis* (Washington, D.C., The World Bank):. 206–207.

What has been done and what can be done? For many decades national governments, multilateral funding agencies, non-governmental organizations and private businesses have designed and implemented development projects at the local and regional levels in order to promote the economic and social development of indigenous communities. Whilst ILO's Convention 169 states in article 7.1: "The peoples concerned shall have the right to decide their own priorities for the process of development as it affects their lives, beliefs, institutions and spiritual well-being and the lands they occupy or otherwise use, and to exercise control, to the extent possible, over their own economic, social and cultural development," unfortunately, due to multiple reasons, this does not always occur. In September 2003 a Korean farmer killed himself in front of the posh convention center in Cancun, Mexico, where the World Trade Organization was deciding the fate of hundreds of millions of poor peasants—among them most of the world's indigenous peoples–. The unrestricted tearing down of tariffs on agricultural and food products demanded by the leaders of the most powerful economies, together with continued high subsidies that rich countries pay their own farmers, has sentenced these millions of poverty-stricken farmers in the poor countries to a slow death. Unless the principles of Convention 169 are adhered to and implemented, the condition of poor indigenous farmers the world over will only deteriorate further.

Recent experience has shown that economic growth must go hand in hand with social concerns if the results are to be effective and make a difference in the lives of individuals and communities. A new approach seems to be taking hold in international discourse: human—rights centered sustainable development, meaning that unless development can be shown to improve the livelihoods of people within the framework of the respect for human rights, it will not produce the desired results. This approach may be of particular importance for indigenous peoples whose human rights have frequently been neglected when not actually impaired by traditional economic development approaches.

There is much international debate on these issues, and it is useful to place the human rights concerns of indigenous people into this frame of reference. Particularly relevant to this topic are the reports of the recent sessions of the WGIP (2001–2003), devoted to the right to development and to globalization and their implications for indigenous people. A review of the numerous statements made by government delegates and representatives of indigenous peoples and NGO's at these sessions points to the serious concerns expressed about human rights issues in the process of development.[74]

[74] Cf. E/CN.4/Sub.2/2001/17 and E/CN.4/Sub.2/2003/22.

5.12 Political Representation, Autonomy, Self-Determination

Indigenous self-organization has made considerable progress over the years. From the local level to the regional, national and international levels indigenous peoples' associations have become social and political actors in their own right, as witnessed by their continuing participation in the yearly sessions of the WGIP. They speak with many voices but on the fundamental issues of their human rights, their objectives and their aspirations they are usually in remarkable agreement. In some countries they are now recognized as legitimate partners and interlocutors of governments and other social sectors on the national scene. In other countries the going has been more difficult, their organizations may not be officially recognized and their human right to free association may not be completely respected. To the extent that the rights of indigenous peoples themselves are sometimes neglected and ignored within existing power structures, their organizations and other human rights advocacy associations that take up their cause may also become victims of abuses and be denied adequate protection under the law. Numerous communications to this effect have been addressed over the years to the UN Office of the High Commissioner on Human Rights (UNHCHR), the ILO Commttee of Experts and, among others, the Inter-American Commission of Human Rights.

Beyond respect for their human rights, indigenous organizations also claim the right to political representation qua indigenous peoples at the national level, an issue which may or may not be compatible with existing political structures. More insistent has been the demand for some kind of autonomy, and in a number of countries this has been achieved whereas in others it is not contemplated in current legal arrangements. A case in point is the Constitution of the Philippines which recognizes the right of Muslim and Cordillera peoples to self-determination in the form of autonomy, but the latter are still awaiting the creation of their autonomous region.[75]

One of the more controversial topics surrounding the human rights and fundamental freedoms of indigenous peoples concerns the much debated right of peoples to self-determination. In their statements to international forums indigenous representatives demand the recognition of their right to self-determination as peoples. Equally insistently, some states argue that such a right should not extend to the indigenous. The concept of self-determination is closely linked to the use of the term 'peoples'. There does not appear to be a clear and unequivocal definition of this term in any of the multiple international legal instruments that have been adopted over the last half century nor, for that matter, in national legislation. Without a clear definition that may command a broad consensus, it is not obvious what the debate is really all about. In political science and legal literature the term is usually linked to all the citizens of an existing state, whereas in more

[75] David A. Daoas, 'The rights of the cultural communities in the Philippines', "...Vines that won't Bind...", Proceedings of a Conference held in Chiang Mai, Thailand, 1995, IWGIA Document 80, 97–107, 102–103.

sociological texts the notion of a 'people' refers to certain commonalities, shared identities and identifications.

The principle of the right of peoples to self-determination has been present in international debates for almost a century, and the current claims to this right by indigenous organizations is only the latest instance of its use in the expanding debate about human rights. Whereas some national constitutions do indeed refer to the right of self-determination of indigenous peoples (e.g. Mexico's reformed constitution of 2001), other legislations avoid it, and the controversy relates to the meaning given to the term in both international and national law. Chile's Congress, for example, has voted against several initiatives that would constitutionally recognize the country's indigenous peoples as such. Africa provides another example of conceptual difficulties. In 1981 the Organization of African Unity approved the African Charter on Human and Peoples' Rights, and yet nowhere is the term 'peoples' defined. Specialists continue to debate whether the term should apply only to all citizens of a given State or whether it has other applications as well (such as indigenous peoples). It is this debate which is holding up the adoption of the Declaration on the Rights of Indigenous Peoples in the United Nations. [Since this was written the UN adopted the Declaration in 2007].

Listening to Arhuaco women in Colombia 2004

Chapter 6
Mexico's Unfinished Symphony: The Zapatista Movement (2000)

Abstract The administration of President Ernesto Zedillo, which took office in December 1994, inherited from its predecessor Carlos Salinas de Gortari, the unresolved issue of an armed anti-government uprising in the state of Chiapas. After the shock-waves of the political transition including a serious financial crisis and the severe devaluation of the Mexican peso—had more or less dissipated, public opinion expected the new administration to address this conflict responsibly and competently, as had been promised during the presidential campaign. Indeed, shortly after the handing-over of power contacts between the federal government and the rebels were renewed, leading eventually to the initiation of a formal 'peace dialogue' which in turn resulted in a signed 'Accord' between the parties in 1996. But thereafter further negotiations stalled, new tensions arose, and the dialogue between the Zapatista National Liberation Army (*Ejercito Zapatista de Liberacion Nacional*—EZLN) and the federal government broke off. The Zedillo administration came to its end 6 years later without having accomplished its aim of solving the armed conflict in Chiapas nor, for that matter, any of the major issues which originated the uprising. In a larger perspective, this omission must be chalked up as a major failing of the Zedillo presidency.

Six years after an indigenous uprising in southern Mexico attracted world-wide attention, a new government took power in 2000. The Woodrow Wilson Center (WWC) for International Scholars in Washington DC organized a collective volume on Mexico's Democratic Challenges, to which I was invited to contribute this evaluation of the Zapatista movement at the beginning of the new decade.This text was first published in 2003 as: "Mexico's Unfinished Symphony: the Zapatista Movement", in: Joseph S. Tulchin and Andrew Selee (Eds.): *Mexico's Democratic Challenges* (Boulder, Colo.: Lynne Rienner): 109–126. The permission to republish this text was granted on 20 July by Ms. Laura Logan, Rights and Permissions, Lynne Rienner Publishers, Boulder, CO, USA.

R. Stavenhagen, *The Emergence of Indigenous Peoples*, Texts and Protocols 3,
DOI: 10.1007/978-3-642-34144-1_6, © The Author(s) 2013

During the waning years of the administration, officials used to state frequently that in contrast to the civil wars in Colombia, El Salvador and Guatemala, the violent conflict in Mexico was quite short-lived, while the peace process was dragging on. It is not clear whether this assessment gives more credit to the government than to the Zapatistas, but in my opinion the emphasis is misplaced. The Chiapas uprising is essentially an expression of a deeply-rooted social and political conflict, and the violent or military aspect of the rebellion is not its most essential feature, much less the cause of it. The failure of the Zedillo administration to solve this confrontation during his sexenio results from the fact that the government was mainly concerned with the overtly military aspect of the encounter—a group of (poorly) armed Indian peasants declaring war on the powerful centralized Mexican state—and was much less interested in dealing with the underlying roots of the rebellion. It is likely that if these social causes are not addressed by the new administration of Vicente Fox in a sensible and coherent manner, the conflict will simmer on for a time and its outcome is by no means predetermined.

The origins and background of the Zapatista rebellion are well known (Collier and Lowery Quaratiello 1999; Diaz 1998; Harvey 1998; LeBot 1997; Tello Diaz 1995; Womack 1999). In a nutshell, a highly stratified and hierarchical political and social structure of colonial vintage succeeded in keeping the Indian peasantry in Chiapas marginalized and socially excluded well into the second half of the twentieth century.[1] (Zebadua 1999) Demographic pressure and soil erosion pushed a growing population out of the traditional Indian highlands who joined an increasing flow of poor peasants to claim land in the tropical lowlands—the Lacandon jungle area next to the Guatemalan border. Here they encountered other settlers seeking subsistence and struggled against the interests of no Indian cattle-ranchers and *latifundistas* (large landholders) who were logging the tropical forest and appropriating what used to be considered nationally-owned lands. Land conflicts, peasant organization and periodic violence marked the opening-up of Mexico's last 'frontier' for several decades during the second half of the twentieth century. The development of infra-structure, the construction of several multi-purpose dams and the opening up of vast new oil fields, offered seasonal non-agricultural employment to Indian peasants, and also contributed to the weakening of communal bonds and solidarity. Government agrarian and social policies were unable to provide adequate solutions to these growing problems, even as the traditional corporate patron-client relationship supported by the official ruling PRI (*Party of the Institutional Revolution*), which was often exercised in autocratic manner by the local governors, was challenged increasingly by radical political organizations. Whereas Samuel Ruiz, the Catholic Bishop of San Cristobal de las Casas (the regional hub city in the Indian highlands) promoted his version of Indian

[1] Marginalization and social exclusion are used here as code words to refer to a heavy-handed and persistent system of discrimination, exploitation and opression that has characterized the world of the indigenous in Chiapas for several centuries (Vogt 1969; Wasserstrom 1983).

Theology (a variant of the Theology of Liberation) to which numerous highland communities adhered, on the other hand a number of Protestant denominations made spectacular inroads among the traditionally Catholic population, particularly in the smaller urban centers and the shanty-towns growing up around the larger cities (Tuxtla Gutierrez, the state capital, San Cristobal, Ocosingo, Comitan).[2] In the early 1980s, another ingredient was added to this bubbling cauldron of rival and often conflicting interests: the arrival of a group of radical left-wing militants from northern Mexico who intended to organize revolutionary activity along the lines of similar movements which in the 1960s and 1970s had taken hold in other Latin American countries. Simultaneously, during the 1980s the border region was to become a haven for tens of thousands of Maya peasant refugees fleeing repression and counter-insurgency in Guatemala. The stage was set for the Zapatista uprising in 1994, though public opinion in Mexico was generally unaware of the rising tensions and unrest among the Indian peasantry of Chiapas. The federal government, however, being better informed, nevertheless chose to ignore them at a time when Mexico was negotiating the NAFTA with the United States.

Rather than face the basic issues raised by the rebellion directly the various demands put forth by peasant and Indian organizations as well as the EZLN–, the government adopted the position that the underlying cause of the rebellion was the 'poverty' of the Indians and that development aid and investments would easily turn the situation around. To be sure, Chiapas occupied one of the lowest rungs of all Mexican states on any economic or social development scale. The 1993 United Nations Human Development Report mentions Chiapas as an extreme case of deprivation on the Human Development Index. Less than 40 % of the population of Chiapas is classified as Indian, but the indigenous are placed systematically lower on any development scale than the non-Indian population, as elsewhere in Latin America (Psacharopoulos and Patrinos 1994).

During the Zedillo years, the government was able to mobilize international resources for projects in Chiapas; in one of his reports to the nation, President Zedillo stated that during his administration 76 million US dollars were destined for Chiapas. It is not clear whether this includes the expenditures of the regular state budget or refers to fresh resources specifically negotiated for development in the state, or to public and private investment. At any rate, no independent objective evaluation of the impact of investments and development aid in Chiapas during the period 1994–2000 is available; nor is it known where these funds end up and who benefits from them, if they have been disbursed at all. Scholars and specialists agree, however, that 'throwing money' at Chiapas does not provide a solution to the conflict nor will it make it go away; certainly increased spending in the region if not accompanied by a democratic consensus will probably lead to greater inequities and social tensions.

[2] Evangelical protestantism has made numerous converts in Latin America in recent decades (Stoll 1990).

But before looking at the possibility of a lasting solution, let us recall the main highlights in the process of war and peace in Chiapas.

The country was taken by surprise on the first of January 1994 when a group of armed and masked guerrillas briefly occupied several towns in central Chiapas and their spokesman, identified as subcomandante Marcos, stated that the EZLN, by declaring 'war' on the Salinas government was fighting against 500 years of oppression and injustice. Basta! (Enough!) they cried, and set out in their first public document the objectives of their armed struggle: work, land, housing, food, education, independence, liberty, democracy, justice and peace.[3]

These demands were formulated the same day that NAFTA entered into force, and at the beginning of that fateful year in which presidential elections were to be held. No doubt the significance of the timing of the uprising was not lost on the Zapatistas, and it certainly spoiled the triumphalism with which the Salinas administration was entering its final year in office.[4]

Over the next few days a number of small battles were fought. The federal army overcame its initial surprise and retook the initiative. The Zapatistas withdrew their forces and retreated to the rural municipios in the region known as Las Canadas (the canyons) whence they had emerged. Informed estimates place the number of victims killed in the fighting at around one hundred and fifty, not many by current genocidal standards of mass killings and ethnic cleansings, but enough to alert Mexican public opinion to the seriousness of the situation and the intentions of the revolutionaries.

The Salinas administration was split between hawks and doves. The former demanded massive and quick military action to liquidate the Zapatista movement at once, arguing that their very presence even if not threatening the stability of the country militarily (no other expression of Zapatista military activity occurred in the rest of the country), was nevertheless destabilizing it in political terms. The doves, however, recognized the social background of the uprising and advised prudence and negotiations. It is to the credit of the Salinas administration that it decided to listen to the doves and within 12 days proposed a cease-fire and the beginning of a dialogue with the EZLN. President Salinas named one of his cabinet ministers (Manuel Camacho Solis, an outspoken 'dove' and a frustrated aspirant to succeed Salinas in the presidency) as his personal envoy and negotiator.[5] The Zapatistas accepted the truce and conditions were ripe for a first encounter between the two sides.

The next six and a half years consisted of a series of false starts, mutual misunderstandings, betrayals and disappointments. The full and detailed history of

[3] Primera Declaracion de la Selva Lacandona. Declaracion de Guerra del Ejercito Zapatista en Chiapas (1 January 1994).

[4] Salinas' hand-picked successor, PRI presidential candidate Luis Donaldo Colosio was murdered in March. Presidential elections were held in July and in December Salinas handed over power to Ernesto Zedillo.

[5] Camacho Solis resigned as negotiator in March, after the murder of Colosio. He later broke with the PRI and became an independent contender for the presidency in 2000.

'what really happened' remains to be written (but will probably never be completely disentangled).[6] Let me briefly recount what I feel to be the major milestones in this process:

(1) The Zapatista uprising immediately received broad media coverage both nationally and at the international level—in stark contrast to guerrilla exploits in Mexico during the 1970s and in other countries—and within days a host of human rights non-governmental organizations and assorted associations had descended upon San Cristobal las Casas to establish a 'peace cordon' around the Zapatista strongholds and monitor possible human rights abuses by the federal army and other authorities. The *Mexican National Commission of Human Rights* sent observers into the area. Press and live TV coverage was widely disseminated, and during the first few days of the rebellion subcomandante Marcos, a literate and articulate spokesman, gave interviews, made statements and established contacts with what was to become a vast world network of sympathizers and solidarity with the Zapatista movement. The skillful use of the mass media became a landmark of the Zapatistas again in contrast to other similar movements—to such an extent that people began to talk about an 'internet war'.[7] The Mexican government was increasingly irked by the attention given the Zapatista movement worldwide and willingly adopted the idea of a 'virtual' war rather than a real one. This cavalier attitude was to guide its involvement in the peace dialogue which followed, and probably contributed to its failure, at least so far.

(2) Though having withdrawn from the urban centers they occupied for a few days, the Zapatistas continued to build up contacts and reach out to civil society in Mexico and abroad, and this vision has certainly been one of their strengths. While official public media at first labeled the Zapatista movement as manipulated by foreign and anti-national interests, made up of 'subversives' and 'bandits' and continued to brand the EZLN as a band of spoilers at a time when Mexico had become a 'first world' country,[8] in general the mass media painted a favorable picture of the movement, underlining its indigenous membership—despite the fact that Marcos was not an Indian—and the legitimate grievances and demands that fueled the uprising. Opinion polls in Mexico during the first few months after the uprising showed a generally favorable attitude towards the Zapatistas among the Mexican people. The EZLN, in turn has continued to court public opinion in various ways and still

[6] As is so often the case, many people know parts of the story, and only a few people may think they know the whole story, but they probably do not.

[7] Ronfeldt et al. (1998).

[8] After a strenuous public relations campaign by the Salinas government, Mexico was accepted as a member of the OECD (Organization for Economic Cooperation and Development), a club of industrial states, the first 'developing' nation to have achieved this status.

counts 6 years later with significant, though fragmented and diminished support among key sectors of the population.[9]

(3) How indigenous was this Indian uprising in southeast Mexico? Critics pointed to the fact that Marcos was not an Indian to disqualify the movement as a whole. But it soon became apparent that the Zapatista army was indeed made up essentially of Maya Indians, and so were its other leaders. Another debatable issue were the Zapatistas' first public proclamations. They dealt with social and economic issues in general but not specifically with issues that mainly concern Indian peoples. Marcos later explained to French anthropologist Yvon LeBot that in order to gain recognition the Zapatistas wanted to be seen as a force for change at the national level and therefore had to deal with larger issues. He also noticed that he had learned much from the Indian peoples with whom he identified and was now conscious of the fact that indigenous issues were indeed of primary importance. This change of heart—or tactics—became apparent in later developments of the Zapatista movement, but it is clear that worldwide support for the movement continued—and continues—precisely because of its indigenous nature.

(4) Peace conversations began in the following months of 1994, but despite the mediation efforts of Bishop Samuel Ruiz, they did not prosper. Tensions increased when in February 1995 the federal army suddenly advanced from its cease fire line and, without firing a shot, occupied positions that had formerly been tacitly accorded the Zapatistas. Simultaneously, the government announced it had discovered the 'true' identity of Marcos and issued orders for his arrest and of some of his putative co-conspirators. It was widely believed that the February offensive was a failed attempt by the army to capture Marcos, but the government stated that it was no more than an effort to bring the Zapatistas back to the negotiating table.

(5) The deteriorating situation finally shook the national Congress into action. Responding to a proposal submitted by the Executive, the Congress enacted a Law for Dialogue and Peace in Chiapas in which it recognized the EZLN as an armed party to the conflict and established the mechanisms to be implemented for the peace dialogue. With the active participation of the now formally established National Mediation Commission (CONAI) headed by Bishop Samuel Ruiz, serious talks between the two parties began in September 1995. Five months later, in February 1996, they signed the

[9] The EZLN organized a 'Democratic Convention' in their jungle stronghold in August 1994, just after the presidential election, which was attended by several thousand participants from all over the country and abroad. In the following years they organized other similar international gatherings, maintaining a constant flow of visitors to their base communities. While the Mexican government has been accused of selectively and illegally harassing numerous 'observers' from different countries—including arbitrary detentions and deportations—in general, considering the fact that a 'war' had been declared in the country, it was surprisingly willing to allow these contacts to continue and the meetings to take place in rebel territory, thinking perhaps to improve its international image and to coax the peace process along.

Accord of San Andres on Indigenous Culture and Rights, the only negotiated agreement so far between the federal government and the EZLN.

(6) The San Andres process was tumultuous and complex. The Zapatistas invited a host of 'advisers' from all sectors of Mexican society, whereas the government had difficulty in even putting together a small group of independent experts, who as it turned out, were more sympathetic to the Zapatista positions. The agreement on Indigenous Culture and Rights was to be the first of seven topics that the parties had earlier agreed to put on the agenda. The second topic on 'Democracy and Justice' never got off the ground, and the rest were not even broached.

(7) Among other points, the San Andres Accord underlines the need for a new legislative framework to regulate the relations between indigenous peoples and the state; the right to self-determination of indigenous peoples as expressed through territorial autonomy within the national state; and the recognition of indigenous customary law. It was agreed that a proposal to this effect was to be presented jointly by the signatories (federal government and EZLN) to the national Congress.

(8) But this was not to be. Once the Accord had been signed, and further dialogue on the other topics did not prosper, not much happened in the way of implementation, even as local conditions in Chiapas deteriorated with the appearance of paramilitary groups, factional strife within indigenous communities and sporadic violence. The government did not appear to be overeager to act on its agreement and the Zapatistas lost faith in the process. In September 1996, a Commission of the National Congress, made up of delegates of every represented political party and known as COCOPA, decided to embark on a 'parallel track' to rescue the peace dialogue. It negotiated discreetly and separately with both parties and in December came up with a draft proposal of a document to be submitted jointly to the Congress itself, which included the essential points of the San Andres Accord. Whereas the Zapatistas agreed to the draft, the government after a few weeks of procrastination decided to propose amendments, which were then rejected by the Zapatistas in January 1997. COCOPA's efforts to bring the sides together had failed and the peace dialogue had been brusquely interrupted.

(9) The government's rejection of a proposal to which it had earlier affixed its signature surprised many observers and exposed the internal contradictions among the country's governing elite. Official statements now argued that the autonomy and the recognition of the customary law of the indigenous peoples was unacceptable and represented a grave danger to national sovereignty. As these were obviously points of some importance to the Zapatistas (as they were, indeed, to numerous indigenous organizations and their advocates), the EZLN felt betrayed once more and withdrew from any further open dialogue with the government, which in turn accused the former of 'intransigence' and unwilling to negotiate. The year 1997 was heavy with foreboding because it now became clear that the peace dialogue was going nowhere, and that neither the government nor the Zapatistas were willing to

retreat from their positions. One side accused the government of not wanting to comply with the San Andres Accord it had signed; whereas the other implied that the Zapatistas never really wanted to reach an agreement anyway. The storm broke dramatically in December 1997, when a paramilitary group of local Indians in the highlands, massacred 47 unarmed pro-Zapatista internal refugees at a prayer meeting—including men, women and children. This was no military encounter between the federal army and the forces of the EZLN, who were holed-up several hundred miles away, but rather a settling of scores between political factions in the highland communities, which had become involved in the struggles over resources and power that was splitting up many of these formerly well integrated Indian societies. The massacre, which drew international condemnation by human rights groups, also exposed another darker side of the conflict: the arming of paramilitary groups by government authorities to divide Indian communities and weaken support for the Zapatistas in the region, and the murkier power politics of local political bosses and caciques. Human rights organizations spoke of the government's low intensity warfare and time-tested counterinsurgency action against the Zapatistas, who in turn denounced the government's 'genocidal' intentions. A war of reciprocal accusations now supplemented the 'netwar', but there was no denying the violence, the fear, the atmosphere of repression that hung in the air around the Indian communities.

(10) By 1998 the peace process was moribund, at least as far as the Zedillo administration was concerned. Though the president shuffled around his cabinet ministers and 'peace negotiators' and publicly invited the Zapatistas at least twice to 'come back to the negotiating table' (in early 1998 and in mid-1999), observers did not see any significant change in the federal government's position on the San Andres Accords. The Zapatistas, in turn, insisted that they would only resume the dialogue if certain previous conditions were met, including a partial withdrawal of the federal army to positions held before the February 1995 offensive, the dismantling of the flourishing paramilitary groups (the attorney general's office had identified at least eighteen such groups), and strict adherence to the signed Accord of San Andres. To make matters worse, in 1998 Bishop Ruiz's *National Mediation Commission* (CONAI) dissolved itself because it felt it had no longer a constructive role to play and the legislative Commission COCOPA was internally divided along political party lines thus effectively neutralized. This is the legacy that President Vicente Fox is now obliged to pick up.

Beyond the details of a tortuous process which has not led anywhere, there are a number of issues at stake which need to be considered. The Zapatistas insist that their fundamental demands be addressed by concrete government actions (the implementation of the San Andres Accords). The government, however, is more concerned with ending the 'state of war' and returning to normalcy. A normalcy that the Zapatistas and many others in Mexico consider a return to the status quo ante, that is, to the situation which impelled the Zapatistas to stage their rebellion in the first place. The Zapatista movement now represents more than a focus of guerrillas threatening the stability of a democratic state (a view derived from earlier guerrilla experiences in other Latin American countries). It challenges the international system (globalization, neoliberalism) on which the Mexican national state nowadays attempts to base its legitimacy. Thus it has been labeled an anti-systemic movement in contrast to other military political uprisings that intend to overthrow an existing government. While Mexico's Octavio Paz at first decried the Zapatista movement as a return to a 'pre-modern' form of struggle, Carlos Fuentes has greeted it as the first 'postmodern' conflict. In fact, however, it is neither, and can best be described as a fully 'modern' type of social movement because the demands it raises are based on the modern—and eternal—principles of justice, equality, dignity, liberty and human rights. By wishing to deny the Zapatistas any kind of political legitimation (which explains the de facto rejection of the San Andres Accord) the Mexican government not only deprives the movement as such of a role in national politics, but it effectively denies the indigenous peoples of the country participation in national affairs—notwithstanding official rhetoric to the contrary.[10]

The EZLN evidently craves a national role beyond institutionalized political party and electoral mechanisms. Numerous indigenous organizations have acknowledged that the position of the EZLN has strengthened their own hand in negotiating an effective space for political action on the national scene, even if they do not support the EZLN's choice of violent action to achieve their ends. There is no doubt that thanks to the Zapatistas, indigenous peoples now command more respect in the country than they have ever enjoyed before. Therefore they reject the government's efforts to reduce the Zapatista movement to mere local significance (not national import), to reduce the various expressions of violence (paramilitary groups, repressive measures on Indian communities by the army or government authorities) to 'intra- or inter-communal rivalry' (as official documents label it), and to deal with indigenous demands as nothing that cannot be processed through traditional time-tested clientelistic channels.

As the Zedillo presidency drew to an end there was much speculation in Mexico as to why his administration had been unable (or unwilling) to solve the conflict in Chiapas. A few months before leaving office, on one of his trips abroad, president Zedillo declared that the Zapatista rebellion was a mere 'incident' in Mexico's history, of no great import to the country. On other occasions he used to say that this was a 'light guerrilla', presumably in contrast to the 'heavy' rebellions and civil wars that decimated other Latin American countries during several decades in

[10] Official opinion holds that the Zapatistas should lay down their arms, take off their masks and transform themselves into a political party. The Zapatistas, however, insist that they will only do this after the peace agreement has been fully implemented. In other words, they feel that only by obtaining the government's compliance to the agreement will they be able to achieve the political legitimacy needed to become an alternative political force in Mexico. For this they count on the continued support of the Mexican 'civil society'—which the government considers a useless distraction to any future peace negotiations.

At an information meeting in Guerrero, Mexico. *Source* Personal photo collection of the author

the twentieth century. Perhaps these outspoken words express a much deeper misunderstanding by the country's governing elites of what the Zapatista movement is all about, a misunderstanding which is likely to be shared by the incoming Fox administration.

As all social conflicts, this one has a number of underlying causes, various plots and sub-plots and a number of possible solutions. There is certainly not an easy way out because the conflict -which in reality subsumes a number of different conflicts—does not have a single cause nor does it have only one possible solution. Moreover, as in Kurozawa's famous film Rashomon, various narratives have been woven about this conflict, which does not only confound public opinion but also specialists and perhaps even the contenders themselves. The conflict is constantly being redefined by the participants and the observers, according to the perspective from which it is looked at and the narratives with which the parameters of the conflict itself are constructed. It is unlikely that it will be resolved unless the quarrel over the definition of the conflict and its various narratives is decided. These issues might be looked at from three different perspectives:

Firstly we must notice the structural conflict. This occurs not only in Chiapas but also in other parts of the country in which there are indigenous peoples and communities. This structural conflict is as old as the social and economic system that produced the great inequalities in economic welfare, social status and political power between the Indian peasantry and the non-Indian population which usually occupies higher positions in the stratification scale. The earlier internal colonialism that prevailed for centuries has been replaced by a postcolonial situation which is more flexible and fluid, and in the roiling waters of modernization Indian communities have become more divided and polarized. The earlier vision of integrated, harmonious communities -as described by anthropologists—a few decades ago, no longer holds. But, conscious of this social disintegration which has many facets, indigenous organizations are pleading for strategies to 'recompose' their communities, and they see the implementation of the San Andres Accords as one way to achieve this objective. Contrary to some currently fashionable assessments, the NAFTA and neoliberal globalization have not led to an across-the-board improvement of living conditions for Indians in Chiapas or elsewhere in the country.[11] Rather, these changes in so far as they affect indigenous communities directly or indirectly, generate growing socio-economic inequalities, the disappearance of solidarity networks in rural villages and townships, the weakening of social compensatory institutions and the growth not only of poverty but of pauperism. While this is not a new phenomenon in world capitalism—once labeled the 'development of underdevelopment'—it is still a major cause of tension and structural conflict.

There are several actors in this conflict: firstly, the indigenous peasantry who have always been the historical victims of this situation. But there are others: local and regional interest groups— landowners, cattle ranchers, merchants large and small and intermediaries; also government bureaucrats and not a few indigenous power-brokers who now have a stake in the system.

Structural conflict is not only a 'class struggle' in the traditional sense of the haves and the have-nots, the privileged and the deprived. In Chiapas it is also en ethnic confrontation between Indians and *mestizos*, which is also deeply rooted in the history of internal colonialism. Whereas some commentators on the Chiapas scene have blamed the anthropologists for inventing ethnic differences, suggesting that if only academics would stop concerning themselves with these topics, people would easily learn to get along together, fact is that ethnic distinctions are deeply rooted in the local imagery as a result of the asymmetrical power relations in the economic and social arenas ever since colonial times. Local people distinguish clearly between Indians, *caxlanes*, *gente de razon* or *ladino*s, which are not only descriptive labels in everyday discourse but concepts used in social and cultural mapping. They refer to different social statuses, systems of interpersonal relations, types of discrimination and exclusion. This is all part of the structural conflict in Chiapas, as elsewhere in the country, and it should not be forgotten that the situation has led to persistent human rights abuses of Indians and peasants, social activists, women, children, migrant workers and settlers, and entire communities; violations that have been assiduously documented by human rights organizations and which surely may be considered as one of the triggering factors of the 1994 uprising.

---

[11] A recent FAO study reports that 40 % of Mexico's population is undernourished.

The structural conflict cannot be solved in the short term; the San Andres Accord or any other negotiated agreement will not do away with it. The social, economic and political structures of inequality will change progressively only by means of a long-term process and persistent social and economic policies that might directly benefit the indigenous peoples and communities through processes of redistribution of power and wealth. Mere legislative changes at the local state or national levels will be necessary but not sufficient to effect such transformations.

The second perspective must focus on the political conflict between, on one side, a political-military organization, the Zapatista National Liberation Army and its non-military members (known as *bases de apoyo*) and other supporters in specific areas of the region, as well as a wide array of peasant and popular organizations, and on the other the local power structure concentrated around the ruling elite of the state of Chiapas. This has traditionally been an autocratic, authoritarian, centralized and antidemocratic structure which at times may appear to be legal and institutional. While some observers accuse the EZLN of being authoritarian (and as all tightly-knit revolutionary organizations it may certainly have an authoritarian streak to it), the real authoritarianism is represented by those who have wielded power against the interests and well-being of the peasants and the Indians.

The main actors in this political conflict which now has also become a military or pseudo-military one, are the peasant and indigenous organizations that address their grievances and demands to the government since their emergence in the 1950s after the Instituto Nacional Indigenista (INI, a federal government agency) established a base of operations in San Cristobal las Casas. These various organizations that multiply and grow during the following decades, are the principal contenders in the political conflict that erupts in the 1970s. Besides the peasant and workers' unions, mention must be made of the religious groups: the Indian theology promoted by Bishop Samuel Ruiz in the San Cristobal Diocese since the 1960s and the various Protestant denominations. An important ingredient was added by the arrival of small organizations of militant leftists, representing different tendencies and who were often at loggerheads with each other, who joined and occasionally managed to lead the struggles of the Indian peasant associations in a pattern of changing alliances. Out of one of these groups came the Frente Zapatista which decided on armed struggle.[12] On the opposite side we find the business and landowners associations, the *autenticos colonos* (local aristocracy) of San Cristobal las Casas, as well as the municipal caciques (political bosses) who are the backbone of the national PRI at the local level. In earlier years, members of some of these sectors fostered the creation of small private armed bands in the service of certain landowners (*guardias blancas*), who used violence and the threat of violence against the leadership and members of popular organizations, often in connivance with government authorities. The picture becomes more complex in the 1980s when new political parties challenge the traditional control of the PRI

[12] See Diaz (1998) and Harvey (1998).

through electoral contests at the municipal and state levels. Inasmuch as the EZLN declared 'war' on the Mexican state, which led to a massive increase of military presence in the area (some estimates place the number of federal troops in the state at 40,000), it is also necessary to take into consideration the various vested interests in the conflict that the military itself has acquired. Finally, one can only speculate about the influence—often mentioned in the media—of drug trafficking on the political balance in the state.

As can be seen, there are numerous different actors with special interests in this political-military conflict, which in turn results from the structural tensions mentioned earlier. What solution(s) might there be to this problematique? The government has often suggested the need for modernization and democratization, but it has been very slow in promoting these tendencies. The traditional power structures in Chiapas are arrayed against the modernization and democratization of political life. This may finally have changed in August 2000, when an alliance of opposition parties was able to defeat the PRI government-supported candidate for governor of Chiapas. The new governor, Pablo Salazar, formerly a member of the PRI, was also a member of the COCOPA, the federal legislative commission dealing with the peace process. With two opposition candidates in the national presidency and the local governor's office, things may finally begin to change in Chiapas.

There is also a third perspective which can be brought to bear on the problem, and this is the fact that since 1994 there has been an armed conflict between the EZLN and the national and state governments. The former has its *bases de apoyo* and supporters and sympathizers in the country and abroad. The government controls the resources, the military power, the public administration and the political institutions. Let us recall that in 1995 the national Congress adopted a law for dialogue and peace in Chiapas in which it legally recognized a 'group of dissatisfied, mainly indigenous Mexicans', in other words, it recognized the EZLN as a de facto if not de jure belligerent in this war. While the duration of open armed conflict was truly short, the period of negotiations has been unduly long. And whereas the number of direct casualties of the fighting was relatively small, the situation has become more complex due to the presence of a number of paramilitary outfits (which the government prefers to call 'armed civilian groups'). As in other areas, the danger these units represent is that in time they may escape from the control of their masters: those who armed and financed them. However, it is not publicly known who in fact is behind these groups, and the federal army vehemently denies any involvement. Nevertheless, they are responsible for numerous instances of violence including murders, torture and abductions of Zapatista sympathizers and presumed members of the EZLN. Their existence and relative freedom of action is an additional factor which makes the resumption of peace talks difficult.

Again, the principal actors in this military conflict are the EZLN and the local army and police units; the political supporters of the Zapatistas such as the Autonomous Municipios, established in defiance of local legislation, the different levels of public administration and both COCOPA and CONAI (despite the fact that the latter has been formally disbanded). The solution to this conflict can only come through continued dialogue and negotiations, though events during the last 3 years hardly lead to optimism.

From the vantage point of this third perspective, it seems to me that there has been an overlap of timing and strategies between the structural conflict and its underlying causes, the political conflict due to incomplete democratization and the armed conflict since January 1994 and everything which occurred thereafter. The three perspectives are linked but must be kept separate.

The Mexican government has occasionally announced major new investments in Chiapas, provided by multilateral agencies and transnational corporations. These are intended to create jobs and improve living standards of the local population, thus helping solve the 'structural' causes of the uprising, referred to above. Where are these resources and what have they accomplished? A recent study finds that between 1994 and 2000 direct foreign investment in Chiapas (on which current development policies rely heavily) amounted to 5.4 million dollars, that is to say less than a million dollars a year. Most of these enterprises are registered in the major cities so it is not known what effects they may have had on incomes in the rural areas, but probably their overall impact is slight.[13]

On the other hand, it has often been said that the EZLN does not trust the political party system and does not believe that democratic elections will alter the system, a belief held by many observers. The Zapatistas have announced repeatedly that they will submit to the larger organization of civil society. But where is this organization? Not much has resulted from the numerous conclaves which the Zapatistas have sponsored in their territory. The PRD, the party sometimes believed to be closest to the EZLN, came in third place in the 2000 presidential elections at the national level, having lost electoral support since 1988. The historical alternative for the country, which the Zapatistas have promised, has not been spelled out clearly and is not reflected in any of the major political platforms. Civil society, despite its diverse manifestations and its strong human rights commitments has not proven adept or organized enough to impose peace on the two contenders; and in the meantime social and economic conditions deteriorate in the area. The real victims of the struggle are the Indian and peasant communities whose situation has deteriorated over the last 6 years. The government has promised more democracy and development, while it has been unable to negotiate a solution to the armed conflict. Between 1994 and 1999 the state of Chiapas has had five governors, none of whom was elected democratically. Perhaps the election of the opposition candidate in 2000 opens a window of opportunity.

Chiapas is Mexico's biggest and most important 'unfinished business'. It expresses the deep social contradictions that national development policies have generated over the last four decades or so (aggravated enormously by neoliberal globalization). Indians (15 % of the Mexican population) have been historically

---

[13] See "Chiapas al dia", in: *Boletin*, No. 216, CIEPAC (29 September 2000); at: < ciepac@laneta.apc.org > .

short-changed. Economic development has passed them by. Despite official rhetoric to the contrary, Indians are the eternally excluded in social, economic and political terms. The Zedillo administration dismissed the Chiapas conflict as a minor 'historical incident', and was unable to offer a satisfactory solution. Hopefully, the Fox administration will do better.

But the Zapatistas have not provided a clear alternative political strategy that may find large-scale echo in civil society: rather, they appear to be carving out a niche for themselves (with difficulty), and to prepare for a 'long march' in political terms. They have not commented publicly on Mexico's political transition, but perhaps they are willing to give the incoming government more of an opportunity to negotiate a peace agreement than the previous one, providing of course that the incoming government will do the same.

There are many points on the agenda for peace, including the definition of new legislation concerning the status of indigenous peoples, their territories, the use of natural resources, the legal recognition of customary law, the rights of the indigenous to political representation, access to the media, and of course the question of self-determination and autonomy. All of which was discussed in the San Andres negotiations and signed into an Agreement.

Beyond the still to be debated issues of democratization, justice and economic and social development for indigenous peoples, there are deeper underlying unfinished themes: human dignity, collective identity, political recognition, social equality and human rights. Mexico's process of democratization is surely incomplete without the participation of the Zapatistas (and more inclusively, of the country's indigenous peoples). It can continue to ignore indigenous demands (as it has over the decades), but it cannot claim to be a truly democratic society until Indian peoples are included in the political and social agenda of the twenty-first century. And this means first of all making peace in Chiapas.

References

Collier, George; Lowery Quaratiello, Elizabeth, 1999: *Basta! Land and the Zapatista Rebellion in Chiapas* (Oakland, Cal.: Food First Books).

Diaz Legorreta, Carme, 1998: *Religion, politica y guerrilla en Las Canadas de la Selva Lacandona* (Mexico D.F.: Ediciones Cal y arenam).

Harvey, Neil, 1998: *The Chiapas Rebellion: The struggle for land and democracy* (Durham, NC: Duke University Press).

Le Bot, Yvon, 1997: *Subcomandante Marcos. El Sueno Zapatista* (Mexico, Plaza & Janes).

Psacharopoulos, George; Patrinos, Harry (Eds.), 1994: *Indigenous people and poverty in Latin America: an empirical analysis* (Washington, D.C.: World Bank).

Ronfeldt, David; et al., 1998: *The Zapatista 'social netwar' in Mexico* (Santa Monica, CA: Rand).

Stoll, David, 1990: *Is Latin America turning Protestant? The politics of evangelical growth* (Berkeley: University of California Press).

Tello Diaz, Carlos, 1995: *La rebelion de las Canadas* (Mexico, Cal y Arena).

Vogt, Evon Z., 1969: *Zinacantan: A Maya community in the highlands of Chiapas* (Cambridge, Harvard University Press).

Wasserstrom, Robert, 1983: *Class and society in central Chiapas* (Berkeley: University of California Press).
Womack, Jr., John (Ed.), 1999: *Rebellion in Chiapas: an historical reader* (New York: New Press).
Zebadua, Emilio, 1999: *Breve historia de Chiapas* (Mexico: Colegio de Mexico and Fondo de Cultura Economica).

Chapter 7
Struggle and Resistance: Mexico's Indians in Transition (2006)

Abstract In 2006 a new administration was elected in Mexico and the Woodrow Wilson Center (WWC) once again gathered a group of scholars to analyze the new political situation. This time I contributed a paper, bringing the story of the Indian movement in Mexico up to the middle of the first decade of the new millennium (This chapter was first published in 2010 in: Andrew Selee and Jacqueline Peschard (Eds.): Mexico's Democratic Challenges. Politics, Government and Society (Washington D.C., Woodrow Wilson Center Press—Stanford, California (Stanford: Stanford University): 251–267. The permission to republish this text was granted on 20 July by the Woodrow Wilson Center Press.).

7.1 From the Zapatista Uprising to Constitutional Reform

Mexico's Indians have been drawn, as other sectors of the nation, into the conflictive politics of what has been called, perhaps over-optimistically, the democratic transition. How has this transition affected indigenous peoples and communities in the country? What role have the indigenous people and their organizations played in this process? What can they expect from, and contribute to, the transformations that have re-energized the political establishment and shaken up civil society?

I will take up the thread of the story from the first year of President Vicente Fox's administration, when he sent a bill to the Senate on indigenous rights and culture. This proposal had been drafted originally by the *Comisión de Concordia y Pacificación* (COCOPA) of the national Congress during the previous administration of Ernesto Zedillo, as a result of the painstaking negotiations in 1995–1996 between the federal government and the *Zapatista National Liberation Army* (EZLN), which produced the San Andrés Peace Accord on indigenous rights and culture. The COCOPA draft was shelved by the government, whereupon the

R. Stavenhagen, *The Emergence of Indigenous Peoples*, Texts and Protocols 3, DOI: 10.1007/978-3-642-34144-1_7, © The Author(s) 2013

Zapatistas broke off all contact with Zedillo. His successor, Vicente Fox, had glibly promised during his campaign that he would solve the Chiapas conflict within 'fifteen minutes', and without further consultation or public debate, in December 2000 he submitted his government's initiative. Despite the national mobilization of the Zapatistas and numerous other indigenous organizations who staged a widely publicized March on Mexico City and whose representatives were allowed to address the Congress, the Senate made several changes to the President's draft and unanimously adopted a constitutional reform on indigenous issues in July 2001. In the lower house a minority (including members of the three major parties) voted against the proposal.

The new text diverged substantially from the original proposal that the Zapatistas had agreed with, especially regarding the issue of self-determination and autonomy for indigenous peoples, control over natural resources, collective ownership of indigenous lands and other points. The new text of Article 2 of the Constitution recognizes the right of indigenous communities and peoples to self-determination, understood as autonomy within the national framework, and leaves it up to the states to legislate to whom this right applies.

The right to self-determination implies the freedom to decide on internal forms of social, economic, political and cultural organization, the application of customary law to the solution of local conflicts, and local self-government according to traditional practices.

The text in fact limits these rights by subordinating them to existing legislation. With regard to the highly controversial land issue, the constitutional text simply states that indigenous communities should have preferential use of natural resources on their lands, within the framework of exiting land tenure legislation, but not the right to own and control these resources. The right to land of indigenous people (a worldwide indigenous claim in recent decades) is subordinated to the prior right of 'third parties'. By hastily adopting the new constitutional text without extensive hearings and debates, the Congress missed a chance to address the major human rights claims of indigenous peoples and to contribute to the peaceful political solution of the conflict in Chiapas. Once again the Zapatistas felt betrayed and withdrew to their haven in the southern mountains.

President Fox did not lobby the Congress for the passage of the bill—a harbinger of the administration's difficult relations with the legislature in the ensuing years—and the final text was the result of negotiations between a handful of senators with hardly any participation by the interested stakeholders, foremost the indigenous organizations. The ensuing ratification process by the states was marred by procedural irregularities which de-legitimized the whole process in the eyes of many. Ten of Mexico's 32 federal states, those with a majority of the country's indigenous population, failed to ratify the constitutional reform. The discontent did not stop there, because shortly after its adoption 331 indigenous municipios asked the Supreme Court to declare the reform unconstitutional. This was surely the first such suit ever brought before the country's highest tribunal. The Court wiggled out of its responsibility by arguing that it had no authority to question the constitutional activity of the legislative branch of the government. It did not judge the substance of the question: whether the rights of indigenous peoples had actually been protected adequately by the adoption of the constitutional reform. The Court's decision raised an important debate among specialists as to the real functions of the various branches of government. The Zapatistas, for their part, supported by a large majority of the country's organized indigenous movement and human rights organizations, clearly understood the message. None of the three branches of government had been willing to address the major political issue: do indigenous peoples in Mexico (who had been constitutionally recognized as such since the 1992 constitutional reform) have recognized rights pertaining to their collective identities and forms of cultural and social organization? The answer in the new constitutional text remains unclear and ambiguous. The indigenous movement did not expect much more from the Fox administration after that. Several indigenous organizations complained to the *International Labour Organization* (ILO) that the process had violated Mexico's international commitment under ILO's Convention 169 on indigenous and tribal peoples (ratified by Mexico in 1990). The petitioners were particularly unhappy with the government's disregard of its obligation to carry out an extensive and participatory consultation among indigenous communities prior to embarking on the constitutional reform process. The ILO's Commission of Experts decided to accept the case.

Even before the constitutional reform, several of Mexico's federal states had passed local legislation or reformed their own state constitutions incorporating the issue of indigenous rights. After the constitutional reform, other states proceeded to adjust their constitutions to the new federal framework The *Special Rapporteur on the Human Rights and Fundamental Freedoms of Indigenous People* of the *United Nations Commission on Human Rights* recommended in 2003 that the Congress reopen the debate on the constitutional reform. This did not happen during the Fox regime, and I doubt that it will occur during the administration of President Felipe Calderón, which has not shown any particular interest in this topic.

7.2 The State and Indigenous Human Rights

The new text of Article 2 of the Constitution is divided into two sections. Section A contains the principles regarding the legal rights of indigenous peoples, whereas Section B, added by the legislators to the original bill, contains a full program for the development of indigenous communities, which expresses the Mexican state's commitment to this important sector of the population. Some experts have expressed their amazement that a constitutional article devoted to the liberties and rights of Mexico's Indians should now also include the outline of practical programs designed to further the wellbeing of these communities, a task that would best be left to secondary legislation. In fact, ever since the nineteen forties the Mexican State has had a structured policy addressing the needs of indigenous communities, known as indigenismo, which has often been described as paternalistic and which has been seriously challenged in recent decades by the emerging indigenous movement.

Early in his administration, President Fox announced that he would pursue 'a new relationship' with the indigenous peoples of the country. He promptly appointed Marcos Matías, an indigenous anthropologist from Guerrero, who had been active in a regional indigenous organization as new director of the by now stodgy and bureaucratic *Instituto Nacional Indigenista* (INI), created in 1948. He also established an advisory office in his official residence to deal with the demands of the Indians, which was headed by a young dynamic woman of Otomi (Ñhañhu) ancestry who had a business background. Xochitl Gálvez had direct access to the President for the full six years of his administration, enabling her to obtain approval for the funding of numerous infrastructure and community development projects in Indian areas of the country. In 2003 the INI was transformed into a new *National Commission for the Development of Indigenous Peoples* (CDI), headed by Gálvez, which continued the traditional indigenista policies with little change. Its Consultative Council made up principally of selected representatives of each of Mexico's 62 identified indigenous ethno-linguistic groups had little say over the President's policy or his decision-making powers regarding indigenous peoples, and was widely considered to be mere window-dressing for the 'new relationship'.

7.3 Major Issues Confronting Indigenous Communities

Around 70 % of all Indians are peasant farmers, and they supplement their mostly subsistence economy with the production of handicrafts and increasingly as migrant day-laborers in the more fertile and productive areas, in the cities and even in the United States and Canada. Studies on the living conditions of the indigenous peoples point to continuing severe deficiencies in the Indian areas. 80 % of the municipios with indigenous population show a high degree of marginality, in comparison with 50 % for the national average. Illiteracy among the indigenous reaches 33 % compared to less than 10 % nation-wide. One-third of all housing in these municipios lacks running water, as against 16 % nationally; while 5 % of all units lack electricity at the national level, the rate is three times higher in the indigenous municipalities, where more than 40 % of all houses have only a dirt floor (the national rate being less than 14 %). At the end of 2006 the CDI presented a report on its achievements during the six years of the administration. It reported that indigenous illiteracy fell by 2.3 % points between 2000 and 2005, and 3.5 points among indigenous women. Access to health services increased by almost seven points, rising from 20 to 27 %. A million indigenous persons had access to health services, through one Fox's favorite programs, the Popular Insurance (Seguro Popular). Access to piped water rose by over six points, from 64 % in 2000 to 70.5 % in 2005; whereas access to sewers increased by 15 points from 40.5 % to 55.6 %, almost double the national average. The number of indigenous households with electricity rose from 83 to 90 %, an increase almost seven times higher than the national average; whereas living units with only dried earth flooring decreased by six points 38 %.

Considering that human rights violations during the previous administration had been one of the issues frequently raised at the international level, in December 2000 the government signed a technical cooperation agreement with the High Commissioner for Human Rights of the United Nations for the establishment of a UN Human Rights office in Mexico,. The OHCHR put together a research team and produced a full diagnosis of the human rights situation in Mexico that includes a chapter on the human rights of indigenous peoples. The report indicates that the majority of the Indian population lives in the poorest municipalities that have the lowest indicators of human and social development. They are usually poor subsistence farmers on small plots of land in a harsh environment, forced in many cases to emigrate in order to survive. It is in these small rural communities that the cultural identity of the Indians is strongest. Their language, social institutions, spirituality, traditional medicine and knowledge, oral literature, music, plastic arts, textiles and other expressions are linked to what anthropologists have called the 'corporate community'. As a result of demographic pressure and market forces, and not least to government policies, these communities long ago began to change as the consequences of globalization make their way into the remotest areas of the country.

The principal human rights issues facing indigenous peoples in Mexico are related to the agrarian question, which has marked the history of Mexico for centuries. The major problem faced by indigenous peoples in Mexico continues to be access to land and its resources. After the agrarian counter-reform of 1992, which opened the way for the privatization of communally held lands, the government established a program of regularization, certification and individual titling of landholdings in the agrarian sector. Inter-community conflicts over land have continued over the years, and the Administration identified 13 serious conflicts in indigenous communities. Through negotiations between the parties aided by the Ministry of Land Reform and the CDI, these conflicts were eventually solved. A number of conflicts over land rights and land use between communities, landowners, ranchers, developers and transnational corporations, have led to serious human rights violations of indigenous peasants, as well as their leaders and allies. Sometimes state authorities and armed actors (police, military and paramilitary groups, private guards, drug lords and others) become involved. A dramatic instance of such agrarian violence took place in 2002 in Agua Fría, Oaxaca, where 26 members of an indigenous community were murdered by armed individuals from a neighboring village, during a drawn-out conflict over forest resources. The *National Commission of Human Rights* (CNDH) accused the government of inappropriate decisions and inaction in the face of rising tensions over land issues which were not stopped in time before the violence erupted.

Similar situations are reported from other communities in several states. Some of these problems were solved in time, such as the situation of subsistence peasant settlers from various parts of the country in the tropical forest reserve of Montes Azules, which had led to some violent encounters. Others continue festering, often complicated by corruption and the interests of local caciques (politically powerful local bosses). Many of these cases are related to the exploitation of forest resources, which as elsewhere, have become highly valuable in the global marketplace. Tensions also arise over the impact of various kinds of development project in indigenous areas. A recent case in point is the government's decision to build a hydroelectric dam, La Parota, in southern Guerrero, which would seriously affect the livelihoods and the environment of several hundred indigenous families. Whereas the authorities argue that they have foreseen sufficient compensation for the communities that would have to be displaced or would be otherwise affected, strong opposition to the construction of the dam has been voiced by numerous communities and organizations, who criticize the government for not having taken their rights and interests into account. They have been harassed and accused of being subversives, if not terrorists, and local authorities have turned a deaf ear to their demands. I visited the area and met with these organizations in 2006, but despite national and international protest, the government appears determined to continue with its plans. Environmentally related human rights concerns have come up increasingly in Indian areas. Massive deforestation, progressive desertification, soil erosion, water pollution, the destruction of coastal ecological reserves due to unstoppable land speculation for tourist development (as along the Caribbean coast), have become more acute in recent years. In most of the affected areas there are indigenous communities that suffer damage (Mayas, Huaves, Lacandones, Tzeltales, Amuzgos, Tlapanecos, Nahuas, Huicholes, Tepehuanes, among many others, have presented complaints). The global interest in biogenetic resources has now reached Mexico's indigenous refuge areas.

Numerous human rights abuses committed on indigenous people result from political conflicts at the local level, usually related to disputes over municipal government, development programs, technical assistance projects, provision of services or the distribution of subsidies. In poor areas, where external institutional and financial resources are scarce, conflict over local political power becomes abundant, particularly when it is abetted, as so often happens, by outside political and economic actors. The state of Oaxaca, which has a high density of Indians, has become a basket-case. In this state 418 out of a total of 570 municipalities (many of them the size of a local rural community with its adjacent lands), elect their local authorities not according to party lists, but by traditional customary rules (usos y costumbres). Still, electoral results are often challenged, post-electoral conflicts and sometimes violence occur. In 2001 nineteen municipal government offices were seized, five people were killed in local violence, and eighteen new electoral processes had to be arranged. In 2006 a violent social conflict erupted between a coalition of teachers unions and popular organizations (including numerous indigenous associations) on the one hand, and on the other the state government. Instead of seeking a politically negotiated solution, the state authorities opted for police and army intervention against the demonstrators, resulting in massive human rights violations (people killed, disappeared, arbitrarily detained, accused of terrorist activities and so on). A new confrontation occurred in July 2007. The National Commission of Human Rights and several highly respected international human rights organizations (among them Amnesty International in August 2007) reported on these violations and denounced the local authorities responsible for the situation. Many people (including some federal legislators) called for the resignation of the state's PRI governor, and a commission in the national Congress initiated an investigation for his possible impeachment. But so far (August 2007) nothing has happened because of the delicate political balance between the ruling parties (PAN, PRI) in the country. The government states that 'everything is in order', the federal administration is unwilling to intervene in the local political situation (even though there has been sufficient evidence of human rights violations during the conflict), presumably because it needs PRI support for its legislative agenda.

7.4 Indigenous Rights and the Justice System

The human rights of indigenous people are often contested in the sphere of justice. There are multiple cases of ethnic and racial discrimination against Indians in the justice system, beginning with the issue of lack of translation services for Indian defendants or plaintiffs who only speak their own native language, the non-recognition of local customary law by judges and prosecutors, and the absence of public defenders. Presumed offenders are often arbitrarily or illegally detained, contrary to existing laws. Physical abuse and cases of torture, as well as extrajudicial killings, of indigenous detainees have been widely documented by human rights defense organizations. Court cases in which Indians are involved are frequently plagued by irregularities and corruption. Some improvement has taken place over the years, particularly now that there are many more active human rights organizations, some with international connections, which make the formerly widespread brutal impunity of violators of indigenous rights more difficult. Some states have established local 'indigenous courts' to deal specifically with indigenous community affairs. An interesting experiment is taking place since 1995 in the state of Guerrero, where local indigenous communities, tired of the climate of persistent insecurity and lawlessness in their areas, decided to set up a Community Police Force under their own control (rather than the official state security forces), which appears to have worked out to their satisfaction. Nevertheless, state and federal authorities have tried to dismantle this body as illegal, in the course of which numerous human rights abuses were committed. Other indigenous communities in the country have set up similar alternative security organizations to protect themselves.

Numerous Indian and other human rights organizations have been lobbying for the full recognition in the justice system of indigenous customary law, which is recognized in the Constitution. The issue has been much discussed among specialists in recent years, but is still far from being put into practice effectively, particularly because there is opposition from the bench for legal, cultural and political reasons. The traditional positivistic legal position holds that there can be only one law for all Mexicans, a position shared by most judges, prosecutors and attorneys in the formal justice administration. Indigenous organizations and human rights defenders argue, on the contrary, that customary law is rooted in the culture of the communities and it is more effective at the local level when administered by traditional village elders. Anthropologists who have studied the issue have shown that the application of usos y costumbres by local authorities usually tends more towards restitution and the re-establishment of communal harmony rather than the punishment of offenders as established in penal legislation, though physical punishment of offenders is not unknown in indigenous communities. It is often argued, in fact, that the adherence to usos y costumbres may imply the human rights violations of individuals as these are set out in national and international law, and that individual human rights must of course be paramount. In the daily practice of justice in indigenous communities the apparent contradiction between state law and customary law is not always so clear, and recent research shows, rather, the existence of an intermediate level of interlegality, where the social actors draw from both sources to find acceptable solutions to conflicts. It is likely that in the coming years the national justice system must adapt to the new reality of a multicultural country.

Many of the human rights issues in the justice system arise because of the tendency by the State to criminalize social protest activities of indigenous and popular organizations. The arbitrary use and abuse of legal mechanisms to thwart social protest (such as in Oaxaca, Guerrero, Puebla, Veracruz or the state of Mexico in recent years), hits indigenous organizations especially hard because they are usually the weakest link in the system. Field research in various states has uncovered the deplorable situation of indigenous prisoners (especially women) in detention facilities.

7.5 Some Social and Human Development Issues

In general, the socio-economic indicators of indigenous women and children are sub-standard. A field study of 100 indigenous women in Oaxaca reports a total of 209 violations of their sexual and reproductive rights. Indigenous women are more than twice as susceptible to maternal mortality as non-indigenous women. Over half of all indigenous children are undernourished and undersized. A growing cause of concern is the situation of indigenous women and children as seasonal migratory laborers in commercial agriculture, who have little defense against labor abuses and exposure to pollutants and toxic materials. The CDI and some state authorities have begun implementing assistance and health programs for their benefit.

Even before Mexico was constitutionally recognized as a multicultural country, the Ministry of Education had developed a program of specialized indigenous education in public elementary schools in indigenous areas, with several thousand bilingual teachers on the federal payroll. Culturally appropriate teaching materials and methods, including textbooks in native languages circulated widely in the country. But the professional level of young indigenous bilingual teachers fell short of the required standard and the program never received sufficient institutional and financial support to make indigenous education on par with the national school system. During the Fox administration the program serviced over one million elementary school children among 47 indigenous groups, with over 50,000 teachers in 19,000 schools. The federal budget for indigenous schools, however, decreased by more than fifty percent between 2002 and 2005. The Ministry of Education has recognized that the terminal efficiency of bilingual primary education is 73.5 % as compared to an 86.3 % national average. During the Fox years the Congress adopted a national law on indigenous linguistic rights and created the National Institute for Indigenous Languages, which had long been desired by indigenous organizations. The government also established a number of new Intercultural Universities in the country, which have yet to get off the ground. Recent federal anti-discrimination legislation is being increasingly used by complainants before the courts. Indigenous communities have long demanded the right to set up their own local broadcasting stations as a community service. National legislation has been very restrictive in this sense and most of their demands were rejected. Congress adopted a new telecommunications law in 2007 which ignored the rights of indigenous communities. The Supreme Court considered this law as discriminatory (not only against indigenous communities), and sent it back to the Congress for review.

7.6 The Indigenous Movement and its Future

Since the Zapatista uprising in 1994, the human rights of indigenous people in Mexico, and particularly in Chiapas, have been front-page news. Due to the vagaries of national politics, the Zapatista conflict was not solved through political negotiations, even though peace talks took place and a peace accord on the culture and rights of indigenous people was duly signed in 1996. In the following years the government attempted to impose peace on the Zapatistas through the alternative use of force and cooptation, what some observers have defined as low-intensity counterinsurgency tactics. Numerous human rights violations were denounced, international human rights organizations visited the area periodically and reported on the situation, donor agencies supplied resources for the Zapatista communities. The worst case was the massacre of 45 men, women and children in a prayer meeting in 1997, in which local paramilitary groups and some government officials were involved, but the crime was never really investigated and the intellectual authors never brought to justice.

The PAN denounced Zedillo's Chiapas policy while it was in the opposition, but when Fox came to power he did little to stimulate a return to a peace dialogue with the EZLN. To be sure, Fox, as demanded by the Zapatistas, downgraded the military presence in the zone and liberated most of the Zapatista prisoners held in federal detention centers, but the failure of the constitutional reform process (according to the EZLN) (see above), made the revival of the peace negotiations an ever remote possibility. Fox named Luis. H. Alvarez, a trusted PAN elder (who had been a member of the COCOPA, while a senator in the previous legislature), to head the government's office for Dialogue and Peace in Chiapas, for six years. While carrying out this task, Alvarez was unable to establish any formal contact with the Zapatistas and all of his initiatives were rejected by the EZLN. He reduced his activity to visiting indigenous communities in Chiapas on numerous occasions, and disbursing resources and financial support to those that were hostile or indifferent to the message of the Zapatistas, a tactic that the latter decried as 'divisive'. Alvarez was named by President Calderón to head the National Commission for the Development of Indigenous Peoples (CDI), which has returned to the traditional schema of welfare handouts to pliant indigenous communities that proved itself so useful as a form of political control by an authoritarian and paternalistic government in previous decades.

In the areas under its control in the mountains and canyons of eastern Chiapas, the EZLN worked hard to establish a form of local autonomy. Because it did not recognize the official municipal administration, it created alternative forms of autonomous municipal government by the name of 'Juntas de Buen Gobierno'. Some time later there was an attempt to create more viable regional units by uniting these Juntas into larger 'Caracoles'. Based on the principle of full participation by all community members, these self-contained alternative local governments began to take on the official duties of local public administration with extremely limited resources and practically no outside support. With very little reliable information available, it is difficult to evaluate the outcomes of this experiment at self-government outside of the official state structures. Observers have noted that community support is strong for the alternative schools that the Zapatistas have set up locally, with the volunteer help of teachers from other parts of the country and even from abroad. While they formally reject any official government support, it is known that some public resources—for example for public health—do reach some of the Zapatista communities. However, it also appears that emerging internal divisions have taken their toll on Zapatista unity. Overall, during the Fox years the public image of the Zapatistas diminished considerably. While sub-comandante Marcos on occasion still commands the attention of the media at home and abroad, and some international events bring visitors from overseas to meetings in Zapatista communities, the government pays little public attention to the Chiapas situation, and the attention of the public has been overtaken by other events.

The elections of 2006 would have been an opportunity for the Zapatistas to return to the political process. People on the left expected the EZLN to announce its support for the PRD, but this did not happen. On the contrary, the EZLN openly dissociated itself from the electoral process and in June 2005 put out the Sixth Declaration of the Lacandon Jungle in which it announced that it would launch 'Another Campaign' (*La otra campaña*). Indeed, during the subsequent months an EZLN delegation led by subcomandante Marcos traveled unhindered throughout the country meeting with grassroots and popular organizations, among them many

Indian communities, in order to promote an alternative political perspective. Observers of the political scene have argued that this Zapatista strategy diverted votes from the PRD candidate Andrés López Obrador, who lost the presidency through judicial decision by a razor-thin margin.

Whereas 'La Otra Campaña' strongly reiterated Zapatista support for the rights of Indian peoples, the major political parties relegated indigenous issues to second place in their platforms and public pronouncements. None of the major parties that contended for the presidency presented a forceful, unambiguous position on indigenous rights related to the issues that divided the country on the constitutional reform five years before. The indigenous movement was not represented in any important way in the political campaigns. A handful of indigenous candidates ran for seats in Congress –as well as local state legislatures- on the regular party lists and the new Congress elected in 2006 includes a small number of legislators who identify themselves clearly as Indians, as there have been in previous legislatures. Currently, there does not appear to be an 'indigenous agenda' in the country's political institutions. The public excitement generated in the nineties by the Zapatista uprising and the indigenous movement, appears to have diminished, at least temporarily.

The indigenous movement in Mexico, which emerged in the nineteen eighties, has not become a major political actor in the country, despite the fact that indigenous people make up between 12–15 % of the national population (according to varying estimates), and more in the southeastern states with a high indigenous density (Oaxaca, Guerrero, Chiapas and the Yucatán peninsula). Numerous indigenous organizations have come and gone over the years characterized by internal divisions and factional strife, rotating leadership, lack of continuity, weak institutional and financial support, as well as ambiguous relations with state institutions that continuously attempt to co-opt and subvert the independence of the indigenous organizations. For years the PRI managed its own indigenous organizations, whereas others in recent years became sympathetic to the PRD.

The appearance of the EZLN had a profound impact on the indigenous movement, and continues to provide ideological guidance to a good many Indian organizations. The Zapatistas promoted the creation of the *Consejo Nacional Indígena* (CNI) which has affiliated units in various parts of the country and is able to mobilize considerable support for occasional national congresses (as on the occasion of the debates surrounding the constitutional reform in 2001). But it lacks a structured organization and a consistent leadership and it operates more as a loose network. Other national-level organizations state their tactical differences with the Zapatistas while they agree with them on basic principles. For example, *Asamblea Nacional Indígena Plural por la Autonomía* (ANIPA) which has its base in Guerrero, suffered internal divisions and has practically disappeared as a major player in the field. More effective for practical concerns are local associations that work on human rights, agrarian and development issues with the local population, and sometimes receive support from second-tier human rights organizations and national and international donors. They usually face important challenges, such as leadership, representation, institution building, social mobilization, political alliances and, in many cases, harassment and repression by local police and public authorities. The lack of consistency of the indigenous movement is reflected in its weak representation at the national political level and its minor role in the democratic transition.

Over the years, indigenous peoples have expressed their struggles and objectives through various kinds of organization, which justify the use of the concept of an 'indigenous movement' in Mexico. Many communities were involved in the fighting during the Mexican Revolution, and the agrarian legislation of the early twentieth century addressed the basic land needs of most of the indigenous communities in the country. The PRI turned this process into political capital by organizing and subordinating for its own purposes regional and national-level peasant organizations that included numerous indigenous groups. During the PRI regime, the Mexican state, guided by the prevailing assimilationist and integrationist policy with regard to Indian populations, did not set up separate organizations to be identified by their ethnic identities, even though this possibility was frequently raised. Indians were supposed to merge into the peasant class, and therefore their class interests were presumed to coincide with those of other rural laborers. In the nineteen seventies, during the presidency of Luis Echeverría, the government helped establish a number of ethnically based 'Supreme Councils' among various of the larger indigenous groups, and it also supported the organization of several national and regional congresses where Indian ethnic issues were discussed openly for the first time. The authority of traditional community elders that in many localities functioned as a form of parallel local government to the formal municipal structures, began to be challenged by a new leadership of younger indigenous professionals and political activists who became active at the regional, national and international levels. New organizations arose in opposition to the existing structures, and many of them became involved in the wider emerging space of civil society.

7.7 Alternatives to Indigenismo

The indigenist policies carried out by the Mexican state for many decades were based on the principle of the assimilation or integration (the words were used indiscriminately) of the Indians into the mainstream national society. Land reform, community development and linguistic and educational policies were designed to this end, which was indeed achieved to a great extent. Nonetheless, Indian cultures were and are resilient and resistant. In recent decades, the rise of a class of Indian intellectuals and professionals contributed to the debate on the concept of national culture and led to a reformulation of indigenist ideologies. The constitutional reform of 2001 reflects these concerns, as it affirms that Mexico is a multicultural nation and that indigenous peoples have certain cultural rights that the State is bound to respect. With all of its limitations, the constitutional reform has opened a new space for the participation of Indians in the national polity and its various institutions. Indigenous peoples are now taking up this challenge in various ways. The Zapatistas and their indigenous sympathizers have created autonomous municipios under difficult conditions to exercise their right to self-determination on their own terms. Others have opted to participate actively in the electoral process at various levels through existing party structures and the traditional clientelist politics of rural Mexico. At the more ideological level, Indian intellectuals are actively engaged in pursuing their cultural rights, reinventing their Indian ethnic identities and constructing ethnic symbols and myths. A flourishing group of writers in indigenous languages are producing a new Indian literature (in Maya, Nahuatl, Purepecha and other native tongues). Ethnobotany, ethnohistory, ethnolinguistics, ethnomusicology and other pursuits have been taken up by academic institutions where Indian specialists are becoming increasingly active. Slow progress is being made in the field of intercultural and bilingual education, where experiments with new pedagogical methods and models are being tried. More daunting is the building of viable indigenous organizations that are able to link community needs to state structures, build bridges between local leadership and national civil and political movements, and take part actively in the national debates on the reform of the State and the reshaping of the nation. The collective rights of the indigenous peoples of Mexico (their needs and aspirations) cannot be dissociated from national concerns regarding development, democracy and governance. In recent years, this challenge was first encountered by the State during the presidency of Salinas de Gortari, it was burned into the national consciousness by the Zapatista uprising in 1994, and it became part of an interrupted, but nevertheless historically significant political and constitutional debate in 2001.

Chapter 8
Postcriptum: The Stavenhagen Collection of Pre-Hispanic Art

During the forties, when Rodolfo Stavenhagen grew up in a middle-class home in Mexico City, the family environment was enriched by discussions not only between fellow exiles from central Europe, but also significantly by a circle of Mexican artists, writers, and intellectuals. It was in those early years that Rodolfo's parents, Kurt and Lore, discovered and fell in love with the pre-Hispanic art of Mexico, that is, the artworks and artifacts produced by the indigenous peoples of Meso-America before the Spanish conquest of 1521. Over the years they built up an important collection of indigenous Mexican cultural items that provided them, and their many visitors, with a new and exciting vision of life and art before the colonial period. It also inspired Rodolfo's first interest Mexico's indigenous heritage. This collection was donated by the heirs of Kurt Stavenhagen to the National University of Mexico (UNAM) and is now housed in a special gallery at the Museum of Tlatelolco in the center of Mexico City, open to the public. Some of the illustrations in this book come from that collection.

(From the author's Personal Retrospective in Rodolfo Stavenhagen: *A Pioneer in Indigenous Rights*. Springer Briefs on Pioneers in Science and Practice, Vol. 2: 11–21).

I have published my personal account of my parents' collection as: "To Have Lived Then: A Vision of Ancient Mexico Through the Eyes of a Passionate Art Collector", in: Centro Cultural Universitario Tlatelolco de la UNAM (Ed.): V*ivir entonces. Creaciones del Mexico Antiguo—Colección Stavenhagen* (Mexico, D.F: UNAM, 16 November 2011): 369–373; see also: Elia Stavenhagen: "Kurt Stavenhagen and his Collection", in: ibid: 374–377.

R. Stavenhagen, *The Emergence of Indigenous Peoples*, Texts and Protocols 3,
DOI: 10.1007/978-3-642-34144-1_8, © The Author(s) 2013

Kurt Stavenhagen (1899–1984) holding an item from his collection of original indigenous art of Mexico, now displayed at the Tlatelolco Museum of the National University of Mexico (UNAM)

Stavenhagen collection of pre-hispanic art at the Museum of Tlatelolco, National Autonomous University of Mexico (UNAM), Mexico City

A Maya mask, an item from the Stavenhagen collection. *Source* Personal photo collection of the author

An item from the Stavenhagen collection. *Source* Personal photo collection of the author

About the Author

Rodolfo Stavenhagen (born 1932) is a Mexican anthropologist and sociologist. He is professor emeritus at El Colegio de México, one of Mexico's foremost social science institutions. From 2001 to 2008 he was United Nations Special Rapporteur for the Human Rights of Indigenous Peoples. Before that he was Assistant-Director General for Social Sciences at UNESCO, president of the Latin American Faculty of Social Sciences, and he taught at numerous universities in Europe and the Americas. The University of Tromso, Norway, awarded him an honorary doctorate. He has worked on human rights, indigenous peoples, agrarian problems, social development and ethnic conflicts. Among his principal publications are *Social Classes in Agrarian Societies* (1975), *Ethnic Conflicts and the Nation-State* (1995), *The Ethnic Question: Development, Conflicts and Human Rights* (1990), *Derechos humanos y derecho indígena en América Latina* (1989), and *Los pueblos indígens y sus derechos* (2008).

R. Stavenhagen, *The Emergence of Indigenous Peoples*, Texts and Protocols 3,
DOI: 10.1007/978-3-642-34144-1, © The Author(s) 2013

About the Book

This is the second volume published in the *Springer Briefs on Pioneers in Science and Practice* on the occasion of the 80th birthday of Rodolfo Stavenhagen, a distinguished Mexican sociologist and professor emeritus of El Colegio de Mexico. This collection of six essays on *The Emergence of Indigenous Peoples* Rodolfo Stavenhagen wrote between 1965 and 2009. These widely discussed classical texts discuss: Classes, Colonialism and Acculturation (1965); Indigenous Peoples: An Introduction (2009); The Return of the Native: The Indigenous Challenge in Latin America (2002); Indigenous Peoples in Comparative Perspective (2004); Mexico's Unfinished Symphony: The Zapatista Movement (2000); Struggle and Resistance: Mexico's Indians in Transition (2006). This volume discusses the emergence of indigenous peoples as new social and political actors at the national and international level. These texts deal with human rights, especially during the years he served as United Nations special rapporteur on the rights of indigenous peoples.

R. Stavenhagen, *The Emergence of Indigenous Peoples*, Texts and Protocols 3,
DOI: 10.1007/978-3-642-34144-1, © The Author(s) 2013

MIX
Papier aus verantwortungsvollen Quellen
Paper from responsible sources
FSC® C105338
FSC
www.fsc.org

If you have any concerns about our products,
you can contact us on
ProductSafety@springernature.com

In case Publisher is established outside the EU,
the EU authorized representative is:
Springer Nature Customer Service Center GmbH
Europaplatz 3, 69115 Heidelberg, Germany

Printed by Libri Plureos GmbH
in Hamburg, Germany